# SELECTION OF IRRIGATION METHODS FOR AGRICULTURE

PREPARED BY
On-Farm Irrigation Committee
of the
Environmental and Water Resources Institute

AUTHORS
C.M. Burt, A.J. Clemmens, R.Bliesner,
J.L. Merriam, and L. Hardy

1801 ALEXANDER BELL DRIVE
RESTON, VIRGINIA 20191–4400

Abstract: This report provides an overview of the various agricultural irrigation methods. The variations of each general method (surface irrigation, drip/micro irrigation, sprinkler irrigation, and subirrigation) are described. These descriptions are useful to both the novice and experienced irrigation professional. Most importantly, this report explains the capabilities, limitations, institutional considerations, and economic factors of the methods and their variations. These explanations will facilitate the proper selection of irrigation method for any particular circumstance. Readers will find that there is no universally "best" irrigation method, and that the proper method selection will depend upon the crop, climate, economics, water quality, support infrastructure, energy availability, and numerous other factors.

Library of Congress Cataloging-in-Publication Data

Selection of irrigation methods for agriculture: committee report / On-Farm Irrigation Committee, Water Resources Division; authors, C.M. Burt ... [et al.].
    p.    cm.
Includes bibliographical references (p. ).
ISBN 0-7844-0462-3
    1. Irrigation. I. Burt, C.M. II. American Society of Civil Engineers. On-Farm Irrigation Committee.
S613.S45 1999
631.5'87–dc21                                                              99-054568

Any statements expressed in these materials are those of the individual authors and do not necessarily represent the views of ASCE, which takes no responsibility for any statement made herein. No reference made in this publication to any specific method, product, process or service constitutes or implies an endorsement, recommendation, or warranty thereof by ASCE. The materials are for general information only and do not represent a standard of ASCE, nor are they intended as a reference in purchase specifications, contracts, regulations, statutes, or any other legal document.
ASCE makes no representation or warranty of any kind, whether express or implied, concerning the accuracy, completeness, suitability, or utility of any information, apparatus, product, or process discussed in this publication, and assumes no liability therefore. This information should not be used without first securing competent advice with respect to its suitability for any general or specific application. Anyone utilizing this information assumes all liability arising from such use, including but not limited to infringement of any patent or patents. *Photocopies:* Authorization to photocopy material for internal or personal use under circumstances not falling within the fair use provisions of the Copyright Act is granted by ASCE to libraries and other users registered with the Copyright Clearance Center (CCC) Transactional Reporting Service, provided that the base fee of $8.00 per article plus $.50 per page is paid directly to CCC, 222 Rosewood Drive, Danvers, MA 01923. The identification for ASCE Books is 0-7844-0462-3/00/ $8.00 + $.50 per page. Requests for special permission or bulk copying should be addressed to Permissions & Copyright Dept., ASCE.

Copyright © 2000 by the American Society of Civil Engineers, All Rights Reserved.
Library of Congress Catalog Card No: 99-054568
ISBN 0-7844-0462-3
Manufactured in the United States of America.

# SELECTION OF IRRIGATION METHODS FOR AGRICULTURE

C. M. Burt[1], A.J. Clemmens[2], R. Bliesner[3], J. L. Merriam[4], and L. Hardy[5]

## PREFACE

This report was prepared to assist engineers, designers, sales people, irrigation managers, students, and others with related interests to become aware of the advantages and limitations of the many available irrigation methods and their variations. The report will assist in broadening considerations for irrigation method selection before the final choice of irrigation method is made. It is intended to reduce the likelihood of falling into a conventional choice just because that is the way it has been done historically. The use of variations, alternatives, and improvements in water application methods may be stimulated. Changes from existing conditions of crops, energy costs, labor, land values, and water availability may require reconsideration of existing conditions. This manual will be of value to all, including the most widely experienced irrigation specialists and designers. It is not a design handbook, but it does indicate important factors to be considered when selecting and designing irrigation systems.

---

1  Dir., Irrigation Training and Res. Ctr., BioResource and Agricultural Engr. Dept., California Polytechnic State Univ., San Luis Obispo, CA 93407.

2  Dir., U.S. Water Conservation Lab., USDA-ARS, 4331 E. Broadway, Phoenix, AZ 85040.

3  Consulting Engr., Keller-Bliesner Engr., 78 East Center, Logan, UT 84321.

4  Prof. Emeritus, BioResource and Agricultural Engr. Dept., California Polytechnic State Univ., San Luis Obispo, CA 93407.

5  Consulting Engr., H&R Engr., Inc., 690 Loring Dr. NW, Salem, OR 97304.

# ACKNOWLEDGEMENTS

In addition to the authors listed, the following individuals contributed to the preparation of this report over the span of about 15 years: Glenn Dobbs, Jack Keller, Marshall English, Allan Halderman, DeLynn Hay, Len Ring, Kenneth Solomon, Lyman S. Willardson, William R. Johnston, and Robert Walker.

Final editing of the report was financially supported by the Water Conservation Office of the Mid-Pacific Region of the US Bureau of Reclamation, USDI, and was conducted at the Irrigation Training and Research Center (ITRC) of California Polytechnic State University, San Luis Obispo, California.

# TABLE OF CONTENTS

**CHAPTER 1. INTRODUCTION** .................................................................................................. 1
  EFFICIENCY, UNIFORMITY AND SCHEDULING INDICATORS AND TERMS ........................................ 1
    *Irrigation Efficiency (IE):* .............................................................................................. 1
    *Irrigation Sagacity (IS):* ............................................................................................... 2
    *Application Efficiency (AE):* ........................................................................................ 2
    *Distribution Uniformity, Low Quarter ($DU_{lq}$):* ........................................................... 3
    *Potential Application Efficiency, Low Quarter ($PAE_{lq}$):* .......................................... 3
    *Soil Moisture Deficit (SMD):* ....................................................................................... 4
    *Management Allowed Deficit (MAD):* ........................................................................ 4
  PERFORMANCE EVALUATIONS ............................................................................................. 4
  WATER DELIVERY SCHEDULE TERMS .................................................................................... 5

**CHAPTER 2. IRRIGATION METHOD SELECTION PROCESS** ..................................................... 7
  IDENTIFICATION OF DEVELOPMENT GOALS ........................................................................... 7
  SITE CONDITIONS ................................................................................................................. 9
    *Institutional Considerations:* ..................................................................................... 9
    *Physical Conditions* .................................................................................................. 12
    *Economic Considerations* ........................................................................................ 20
  SYSTEM PRE-SELECTION .................................................................................................... 21
  FEASIBILITY DESIGN AND ECONOMIC ANALYSIS ................................................................. 21
  FINAL SELECTION ............................................................................................................... 25

**CHAPTER 3. SURFACE IRRIGATION** ...................................................................................... 27
  DESCRIPTION ...................................................................................................................... 27
  TYPES OF SURFACE IRRIGATION METHODS ......................................................................... 27
    *Basin* ........................................................................................................................... 27
    *Border Strip* ............................................................................................................... 32
    *Continuous Flood (Basin Paddy)* ............................................................................. 36
    *Ponding (Fill and Drain)* ........................................................................................... 38
    *Furrows* ...................................................................................................................... 39
    *Corrugations* ............................................................................................................. 48
    *Contour Ditches (Wild Flood)* .................................................................................. 49
  CAPABILITIES AND LIMITATIONS .......................................................................................... 49
    *Crops* .......................................................................................................................... 50
    *Soils* ............................................................................................................................ 50
    *Topography* ............................................................................................................... 50
    *Water Supply* ............................................................................................................. 51
    *Salinity/Water Quality* .............................................................................................. 51
    *Climate* ....................................................................................................................... 51
    *Efficiency* ................................................................................................................... 52
    *Irrigation Scheduling* ................................................................................................ 53
  INSTITUTIONAL CONSIDERATIONS ....................................................................................... 54
    *Labor* .......................................................................................................................... 54
    *Service Availability* ................................................................................................... 54
  ECONOMIC FACTORS .......................................................................................................... 55
    *Capital Costs* ............................................................................................................. 55
    *Energy Cost* ............................................................................................................... 58
    *Labor Cost* ................................................................................................................. 58
    *Operation and Maintenance Cost* ........................................................................... 59

# CHAPTER 4. DRIP/MICRO IRRIGATION ........... 61
## DESCRIPTION ........... 61
## TYPES OF DRIP/MICRO IRRIGATION ........... 64
*Orchard/Vineyard Drip (Above Ground)* ........... 65
*Orchard/Vineyard Drip (Subsurface)* ........... 67
*Orchard/Vineyard Microspray (and Microsprinkler)* ........... 68
*Row Crop Drip (Above Ground)* ........... 70
*Row Crop Drip (Subsurface)* ........... 71
## CAPABILITIES AND LIMITATIONS ........... 73
*Crops* ........... 76
*Soils* ........... 77
*Topography* ........... 77
*Water Supply* ........... 77
*Salinity/Water Quality* ........... 77
*Climate* ........... 78
*Efficiency* ........... 78
*Irrigation Scheduling* ........... 79
## INSTITUTIONAL CONSTRAINTS ........... 79
*Labor* ........... 79
*Service Availability* ........... 80
## ECONOMIC FACTORS ........... 80
*General* ........... 80
*Capital Costs* ........... 81
*Energy Costs* ........... 82
*Labor Costs* ........... 82
*Management Costs* ........... 83
*Operation and Maintenance Costs* ........... 83

# CHAPTER 5. SPRINKLER IRRIGATION ........... 85
## DESCRIPTION ........... 85
*Rotating Head Sprinklers* ........... 85
*Low Pressure Spray Nozzles* ........... 87
*Undertree Sprinklers* ........... 88
*Perforated Pipe* ........... 88
## TYPES OF SPRINKLER SYSTEMS ........... 89
*Hand Move Portable / Lateral Move Portable* ........... 89
*End-tow Lateral* ........... 91
*Side Roll / Wheel Line* ........... 92
*Side Move Lateral* ........... 93
*Traveling Gun and Rotating Boom Systems* ........... 94
*Center Pivot Systems* ........... 95
*Linear Move (Lateral Move) Systems* ........... 98
*Solid Set Systems* ........... 100
*Undertree Orchard Sprinkler Systems* ........... 102
*Overtree Orchard Sprinkler Systems* ........... 103
## CAPABILITIES AND LIMITATIONS ........... 104
*Crops* ........... 105
*Soils* ........... 105
*Topography* ........... 105
*Water Supply* ........... 106
*Salinity/Water Quality* ........... 106
*Climate* ........... 106
*Efficiency* ........... 107
*Irrigation Scheduling* ........... 108
## INSTITUTIONAL CONSIDERATIONS ........... 109

  *Labor* ............109
  *Service Availability* ............109
 ECONOMIC FACTORS ............110
  *Capital Cost* ............110
  *Energy Cost* ............110
  *Labor Cost* ............111
  *Management Cost* ............111
  *Operation and Maintenance Costs* ............111
  *Special Cultural Costs* ............112

**CHAPTER 6. SUBIRRIGATION - WATER TABLE MANAGEMENT** ............**113**

 DESCRIPTION ............113
 CAPABILITIES AND LIMITATIONS ............113
  *Crops* ............113
  *Soils* ............113
  *Topography* ............114
  *Water Supply* ............114
  *Salinity / Water Quality* ............114
  *Climate* ............114
  *Efficiency* ............114
 INSTITUTIONAL CONSIDERATIONS ............115
  *Labor* ............115
  *Management Skills* ............115
  *Service Availability* ............115
 ECONOMIC FACTORS ............115
  *Capital Cost* ............115
  *Energy Cost* ............116
  *Labor Cost* ............116
  *Management Cost* ............116
  *Operation and Maintenance Cost* ............116

**REFERENCES** ............**119**

**NOTATION** ............**123**

**INDEX** ............**125**

# LIST OF FIGURES

Figure 2-1. Benefit/Cost Envelope for Economic Analysis of Development Projects..................8

Figure 3-1. Large Scale Basin Irrigation with Corner Inlet. .......................................................28

Figure 3-2. Small Scale, Hand Constructed Bedded Basin Irrigation. ......................................31

Figure 3-3. Border Strip Irrigation with Pulled-up Borders. .....................................................33

Figure 3-4. Small Scale Continuous Flood Irrigation of a Rice Field. ......................................38

Figure 3-5. Alternative Row Furrow Irrigation with Siphon Tube Supply. ...............................41

Figure 3-6. Schematic of a Return Flow System Used in Conjunction with an Underground Distribution Pipeline. ...............................................................................................42

Figure 3-7. Cablegation. ............................................................................................................44

Figure 3-8. Principles of Surge Flow .........................................................................................45

Figure 3-9. Experimental Surge Flow System with Individual Valves at Each Furrow. ..........46

Figure 3-10. Surge Flow System with Single Main Valve Diverting Water to Two Sections of Gated Pipe. ...............................................................................................................46

Figure 4-1. Schematic of Typical Drip/Micro Irrigation System on Trees or Vines..................62

Figure 4-2. Microspray System on a Newly Planted Almond Orchard. ...................................62

Figure 4-3. Schematic of a Subsurface Row Crop Drip Design Typical of the Central Coast of California. ................................................................................................................63

Figure 4-4. Block Valves for a Permanent Subsurface Row Crop Drip Design. ......................63

Figure 4-5. Suspended Hose and Emitter on a Vineyard ..........................................................66

Figure 4-6. Roots Which Were Uprooted During Installation of Buried Drip on Established Pistachio Trees. ........................................................................................................68

Figure 4-7. Microsprayer During Irrigation. .............................................................................69

Figure 4-8. Media Filtration for a 60 ha Row Crop Drip System. ............................................73

Figure 4-9. Fertigation Tanks for Row Crop Drip. ...................................................................75

Figure 5-1. Typical Hand Move Lateral Sprinkler System. ......................................................90

Figure 5-2. Hand Move Lateral Sprinkler System. ...................................................................90

Figure 5-3. Typical Towable Lateral Sprinkler System. ...........................................................91

Figure 5-4. Typical Side Roll Lateral Sprinkler System. ..........................................................93

Figure 5-5. Typical Lateral Sprinkler System. ................................................................................94

Figure 5-6. Typical Traveling Gun Sprinkler. ................................................................................96

Figure 5-7. Typical Rotating Boom Sprinkler. ...............................................................................96

Figure 5-8. Typical Center Pivot Sprinkler System. ......................................................................97

Figure 5-9. Typical Linear Move Sprinkler System with Sprinklers on Booms. ..........................98

Figure 5-10. Typical Linear Move Spinkler System with LEPA Sprinklers. ................................99

Figure 5-11. Typical Linear Move Sprinkler System with LEPA Sprinklers on Onions. ............100

Figure 5-12. Portable Aluminum Solid Set Sprinkler System on Carrots. ...................................101

Figure 5-13 Typical Undertree Orchard Sprinkler System. .........................................................102

# LIST OF TABLES

| | |
|---|---|
| Table 2-1. | Factors Affecting the Selection of Different Types of Modern Irrigation Systems for Use in Developing Countries (Modified from Keller and Bliesner, 1990). ...............11 |
| Table 2-2. | Physical Site Conditions to Consider in Irrigation System Selection........................15 |
| Table 2-3. | Suggested Guide for Selection of Irrigation Method................................................16 |
| Table 2-4. | Typical Economic Lives and Maintenance Costs for Irrigation System Components. ....................................................................................................................................24 |
| Table 2-5. | Average Operating Labor Requirement for Sprinkler and Drip/micro Irrigation Systems. (Add Surface Irrigation Labor Requirements.) ........................................25 |
| Table 3-1. | Surface Irrigation Methods and Variations................................................................28 |
| Table 3-2 | Potential Application Efficiency for Surface Irrigation Methods Under Ideal and Practical Conditions...................................................................................................52 |
| Table 3-3. | Unit Costs for Surface Irrigation Components. .......................................................55 |
| Table 3-4. | Estimated Construction Cost Ranges for Various Surface Irrigation Methods .........56 |
| Table 3-5. | Labor Requirements for Surface Irrigation. .............................................................58 |
| Table 3-6. | Typical Operation and Maintenance Costs for Surface Irrigation Methods ..............60 |
| Table 4-1. | Flow Rates of Typical Emission Devices in Drip/Micro..........................................64 |
| Table 4-2. | Approximate Initial Costs for Drip/Micro Irrigation Systems in the U.S. (Excluding Design)...................................................................................................82 |
| Table 5-1. | Attainable Application Efficiencies ($AE_{lq}$) for Sprinkler Irrigation Systems..........107 |
| Table 5-2. | Operating Labor Requirements for Sprinkler Irrigation Systems...........................109 |
| Table 5-3. | Capital Cost Ranges for Sprinkler Irrigation Systems............................................110 |
| Table 5-4. | Energy Requirements for Sprinkler Irrigation Systems..........................................111 |
| Table 5-5. | Average Annual Maintenance Costs for Sprinkler Irrigation Systems...................112 |

# CHAPTER 1
# INTRODUCTION

Frequently, the selection of an irrigation method (either on new ground or for a modernization program) is made without adequate awareness of the capabilities of the various available methods and systems, the choice resulting in less-than-optimal irrigation. The purpose of this manual is to present to users a process of selecting an irrigation method and to inform them of the capabilities and limitations of the most widely used irrigation methods.

The irrigation method selection process considers the broad categories of (1) institutional, (2) physical, and (3) economic factors. The discussion of each irrigation method is organized to address capabilities and limitations as they relate to these items. If variations of a given method are common, they are also discussed. The irrigation methods are organized under 4 major types: (1) Surface, (2) Drip/Micro, (3) Sprinkler, and (4) Subirrigation. These categories are further divided into the various methods that fall under that particular type.

Some basic terminology and definitions are needed to understand and evaluate irrigation systems. Such terms as efficiency, uniformity and scheduling are pertinent to all irrigation methods and are presented here.

## EFFICIENCY, UNIFORMITY AND SCHEDULING INDICATORS AND TERMS

If meaningful comparisons of system types are to be made, one set of performance indicators and terms must be used for all irrigation methods. Over the years, many performance indicators have been utilized. Some of these indicators, such as Christiansen's coefficient of uniformity for sprinklers, have very specific applications. However, for comparisons between methods and their use, it is essential that a common terminology be used. This manual utilizes the definitions which follow and which are more fully described in the Task Committee Report by the American Society of Civil Engineers On-Farm Irrigation Committee, Irrigation Performance Measures: Efficiency and Uniformity (Burt et al., 1997).

### Irrigation Efficiency (IE)

$$IE = \frac{\text{Volume of irrigation water beneficially used}}{\text{Vol. of irrig. water applied} - \Delta\text{Storage of irrig. water}} \times 100 \qquad (1\text{-}1)$$

A beneficial use of water supports the production of agricultural crops. This term includes the description of processes such as leaching, weather control, and evapotranspiration (ET) from non-crop plants beneficial to the crop (wind breaks or erosion control cover crops), in addition to crop evapotranspiration ($ET_c$).

Determination of irrigation efficiency requires definition of both a boundary and a time interval. Irrigation water moving into the space defined by the boundary (e.g., field, farm, irrigation district or river basin) over a given time interval (e.g., one irrigation cycle, one irrigation season, one year, etc.) becomes the applied volume. Beneficial uses are uses that occur in the region defined by the boundary. If at the end of the time period the irrigation water content within the designated region is the same as it was at the start, $\Delta$ storage = 0. IE may be defined in terms of depth rather than volume, where depth is defined as the total irrigation water volume divided by the area enclosed by the boundary.

IE is only applicable in evaluating the performance of an irrigation system for specific site conditions over some set period of time.

**Irrigation Sagacity (IS)**

$$IS = \frac{\text{Vol. of irrig. water benef. used and/or reasonably used}}{\text{Vol. of irrig. water applied} - \Delta\text{Storage of irrig. water}} \times 100\% \qquad (1\text{-}2)$$

The term, reasonable uses, applies to uses from which benefits accrue to society, such as that required to support riparian wildlife, or may include economically and technically unavoidable losses in an irrigation system. IS is more difficult to quantify that IE, but important in the setting of assessing prudent use of a total irrigation water supply. In most settings the items included in "reasonable uses" will have a political, economic and scientific justification basis.

**Application Efficiency (AE)**

$$AE = \frac{\text{Avg. depth of irrig. water that contributes to a stated target}}{\text{Avg. depth of irrig. water applied}} \times 100\% \qquad (1\text{-}3)$$

The stated "target" may be the soil-moisture deficit (SMD), it may contain a leaching fraction, or it may simply be a target irrigation depth.

AE is useful in a predictive setting since it is measured against a particular requirement without the need to assess beneficial use. Further, a stated requirement does not require a time period to be specified. It also assumes that the requirement is uniform over the area in question. Since AE does not consider the uniformity of application, it is more useful as an irrigation management tool than an evaluation tool in comparing irrigation methods. AE is used for field irrigation, whereas IE and IS may be used for a field, farm, irrigation district, or basin. AE is also used for a single irrigation event, whereas IE and IS can be used for a variety of time intervals.

## Distribution Uniformity, Low Quarter ($DU_{lq}$)

$$DU_{lq} = \frac{d_{lq}}{D_{avg}} = \frac{\text{average low quarter depth}}{\text{avg depth of water accumulated in all elements}} \quad (1\text{-}4)$$

Distribution uniformity is a measure of the uniformity with which irrigation water is distributed to the plants in a field. The practice of using the least watered 25% of the area (low quarter) as the reference standard has gained wide acceptance (Burt et al., 1997). The uniformity described by $DU_{lq}$ (and all terms involving the low quarter) leaves about 1/8 of the area at less than the value of the numerator. This "under irrigation" varies from zero at the 1/8 point to the minimum depth applied at the extreme. This term can be applied to all irrigation methods.

Statistically based expressions of uniformity have historically been used. The Christiansen Uniformity Coefficient (UC) was the first of such methods and has been widely used in the sprinkler industry. For normally distributed data it is equivalent to $DU_{low\ half}$ and is not recommended in making comparisons between irrigation methods. The coefficient of variation (CV) is another statistical expression of water application uniformity requiring a large number of sampling points and has typically been used in the drip/micro irrigation industry to describe one small component of field uniformity - that of manufacturing variation of emitters.

$$CV = \frac{\text{Std. Dev. of accumulated water depths (weighted by area)}}{\text{Mean water depth}} \quad (1\text{-}5)$$

For normally distributed data it has been shown (Hart, 1961; Hart and Reynolds, 1965) that CV is related to $DU_{lq}$ by the following relationship:

$$DU_{lq} = 1 - 1.27\,CV \quad (1\text{-}6)$$

## Potential Application Efficiency, Low Quarter ($PAE_{lq}$)

$$PAE_{lq} = \frac{\text{depth of irrig. water contribut. to } d_{lq}}{\text{depth of irrig water appl. such that } d_{lq} = \text{rqmnt}} \times 100\% \quad (1\text{-}7)$$

$PAE_{lq}$ combines the concept of application efficiency with distribution uniformity to define the highest possible application efficiency for a system with proper management at 100% Adequacy. $PAE_{lq}$ is based on the concept that the application could be terminated at such time that the target would be just met by the average of the lowest values in the irrigation infiltration distribution. It is difficult if not impossible to make this match with some surface irrigation methods under conditions of a low soil moisture deficit and high intake rate soils. $AE_{lq}$ is most useful for pressurized irrigation methods that can easily control the infiltrated depths.

The application efficiency of the low quarter ($AE_{lq}$) can be computed using equation 1-3 with the lesser of $d_{lq}$ and the actual water requirement as the required depth. This allows measurement of the combined effects of design and management on performance.

**Soil Moisture Deficit (SMD)**

SMD is the difference in the depth of water actually stored in the crop root zone at any given time and the depth of water stored in that crop root zone at field capacity.

**Management Allowed Deficit (MAD)**

MAD (Merriam, 1966) is the desired soil moisture deficit at the time of irrigation. It is the value selected by irrigation management that relates to the optimum allowable soil moisture stress for the crop-soil-water-weather system. It is first expressed as the management allowable percent deficient of the available soil water in the root zone with its corresponding stress, and then as the resultant depth deficit to be replaced for a specific soil and root zone. It is the depth used when designing a system.

## PERFORMANCE EVALUATIONS

Field techniques are available for evaluating efficiencies and uniformities of irrigation methods. Pioneering work was done by Merriam and Keller (1978). Clemmens and Dedrick (1981) described techniques for determining DU in basins. ASAE (1990) has Engineering Practices for the evaluation of uniformity for furrow, center pivot/linear sprinkler, and microirrigation systems, although they do not have a standardized concept of evaluating a field for an overall (global) DU. Training classes for professionals on how to do field evaluations have been in progress since 1984 (Burt et. al, 1995) at the Irrigation Training and Research Center (ITRC) in San Luis Obispo, CA, with continuous development and improvement of computerized DU and efficiency evaluation procedures for various irrigation methods.

## WATER DELIVERY SCHEDULE TERMS

It is important to have a common basis for discussing irrigation water delivery schedules by irrigation districts or projects. Water delivery schedules have a large impact on the selection of irrigation methods. These schedules are classified as "demand", "arranged", and "rotation" schedules (On-Farm Irrigation Committee, 1978). Although the discussion below distinguishes between farm and supplier costs, the worldwide trend is towards sustainable water user associations. With these associations, the farm and district become one economic unit because the farmers must pay for costs of (and receive the benefits from) the water delivery system.

The "demand" schedules are flexible as to frequency, rate and duration of irrigation, are user controlled and require no communication system. When the maximum rate is limited to a value that does not appreciably restrict operation conditions, it may be called a "limited rate demand."

"Arranged" schedules require a communication system between the user and the supplier to arrange the conditions of frequency, rate and duration of irrigation. These conditions can result in greater restrictions on farm operations, but less operating or capital costs for the supplier. On-farm reservoirs frequently are beneficial to farms supplied by restricted arranged or rotation schedules. The "limited rate arranged" schedule is a very practical water delivery schedule.

"Rotation" schedules are predetermined and employ rigid delivery conditions. Water is delivered in fixed amounts and fixed intervals. Rotation schedules inhibit best farming and irrigation operations, but are often the least expensive for the supplier. On-farm reservoirs are usually beneficial, and often economical, to farms supplied by rotation schedules (Replogle et al., 1981).

Many other names are used to describe various irrigation schedules, but they are usually colloquial and are not part of a classification system. On-Farm Irrigation Committee (1997) has discussions of these schedules.

# CHAPTER 2
# IRRIGATION METHOD SELECTION PROCESS

Intelligent selection of the irrigation method that best meets the goals and objectives of an irrigated agricultural development is of primary importance in the design process. A system may be well designed but still be an inappropriate selection if another irrigation method would better meet the goals of the development.

While knowledge of the characteristics of the various irrigation methods and the capability to produce optimal designs are essential in the irrigation method selection process, they are not sufficient. The additional necessary step is to select the best irrigation method and configuration of irrigation system for each specific site situation.

To complete this essential function in the selection and design process, an efficient procedure is presented to select the most appropriate irrigation method(s) for meeting the development goals. The process includes five essential steps:

1. Identification of the development goals.

2. Definition of the site conditions (physical and institutional).

3. Pre-screening to select the set of most promising, adaptable irrigation methods.

4. Feasibility level design and economic analysis of the systems from the pre-selected methods.

5. Comparison of the results of step four for each system against the goals defined in step one for final selection.

The selection procedure should include all potential irrigation methods, at least through the pre-screening phase. Too often a valid, and possibly the best, choice is over-looked due to pre-disposition of the designer or farmer to certain irrigation methods.

## IDENTIFICATION OF DEVELOPMENT GOALS

At first glance, it may seem that the obvious goal of all irrigation development is economic efficiency or maximum economic return. While it is true that most irrigation development projects seek economic efficiency, other economic, environmental, and social goals may be considered.

In terms of economic efficiency, a project may focus on either maximum return on investment (maximum benefit/cost ratio) or maximum net benefits from the development. The difference may be best illustrated by examining Figure 2-1. The range of project development options may be described as a continuum through the various development options as depicted by the curved line (B/C continuum) on Figure 2-1. (This is an over-simplification, since the options would typically define a space rather than a line, but the figure is useful for illustration purposes.) The maximum return on investment may be defined as the point at which a straight line drawn through the origin is tangent to the curve. Maximum net benefits occur at the point were a line parallel to the B/C = 1 line is tangent to the continuum. The development continuum usually crosses the B/C = 1 line twice. The space described by these two lines is the economic development space and is the shaded area on Figure 2-1. Any development option falling within the space will have benefits exceeding costs. The economic efficiency may be constrained by social or environmental goals that outweigh the economic goals and compel the selection of a system with less than maximum economic efficiency. In defining the goals of the project, the limits of the constraints of these goals must be identified to properly compare systems.

Other economic goals may also be considered. These goals will have an impact on the economic efficiency and play an important role in the selection process. For example, if capital is limited, a development goal would be to minimize initial cost, even though in a pure economic sense it may be more efficient to increase the initial cost in favor of reduced long term operating costs.

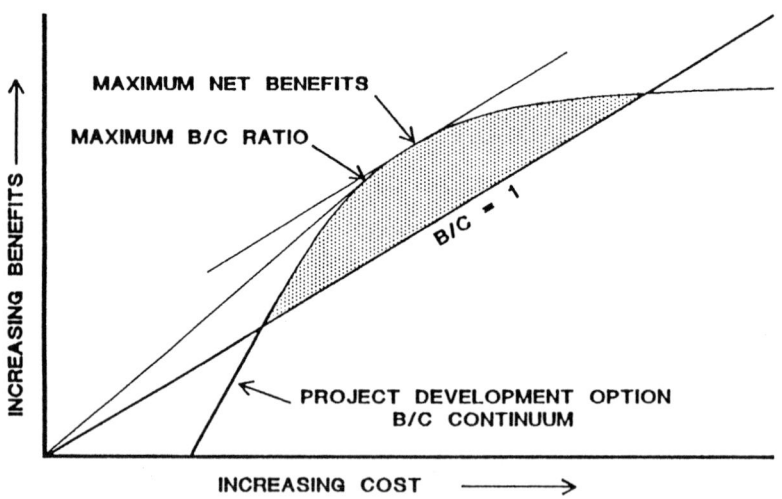

Figure 2-1.    Benefit/Cost Envelope for Economic Analysis of Development Projects.

Social goals are those that would impact some segment of society in ways that may be related to the project development. Some of the more obvious social impacts would include providing jobs for unemployed people, providing locally produced food to limit imports or using locally produced equipment.

Environmental goals usually relate to the impact of irrigation on the environment. Water quality issues may require limitation of runoff or deep percolation from systems. There may be a desire to provide wildlife habitat in conjunction with irrigated agriculture. Any potential environmental relationship that could impact the design of the system, if a particular environmental goal is set, should be considered.

Social and environmental goals may be defined economically by placing a value on the particular goal or may be expressed in terms of potential government subsidies. In these cases, the social and environmental goals do not require separate consideration. However, if these goals are not defined economically, the level of impact allowed on the economic goal must be known as a part of the overall definition of the goal.

## SITE CONDITIONS

Informed irrigation method selection is strongly dependent upon understanding the local site conditions. Knowledge of the physical site conditions is also necessary to the design process. In the selection process, the differential impact to the various system types is important as well as the absolute relationship between site conditions and design considerations for each particular irrigation method. Examples are soil types, slopes, elevations, and existing infrastructure.

In method selection, institutional and economic conditions become equally important with physical conditions. Inadequate understanding or lack of consideration for these conditions could lead to failure of a system that is otherwise physically well designed.

It should be noted that a designer may become familiar with many of these considerations for a particular area in which he works, requiring the detailed information to be gathered only once. However, when working in areas that are unfamiliar, consideration should again be given to each category before final system selection and design takes place.

### Institutional Considerations

Institutional considerations of irrigation system selection are those that relate to people-system interaction. These considerations are often poorly addressed in the selection process and are difficult to quantify. However, since they may over-ride economic or even physical considerations, they should be considered first in the selection process.

Political and legal issues are of primary importance in irrigation system selection. Included are such issues as land use regulations, water rights, containment or disposal of runoff waters, taxation, financial incentives from governments and zoning or construction permit requirements. Before the selection process begins those involved should fully understand the political and legal constraints and incentives.

Since some irrigation methods require a moderate to high level of sophistication in the equipment, availability of support for maintenance is important. The repair capability of the farmer as well as the services available for repair and maintenance must be understood.

The degree of local cooperation required must also be considered. This may include a requirement to share water distribution facilities, agreeing on water delivery schedules, sharing labor, and arriving at policies for control of runoff or return flow.

Although labor cost is often considered an economic issue, it is also an institutional issue in terms of availability and reliability. If adequate labor is not available or dependable, then methods that have reduced labor requirements would be attractive, even if strict economics would suggest greater labor intensity. Understanding the characteristics of the labor pool becomes critical for current and future conditions. This applies not only to availability and reliability, but to level of training as well.

Institutional considerations in irrigation method selection in developing countries have become paramount, especially with modern, more complicated systems. Concerns of special interest in developing nations are described in the summary shown in Table 2-1. Table 2-1 has been revised from an earlier table by Keller and Bliesner (1990), which uses the following definitions:

> Divisibility. Divisibility refers to the suitability of an irrigation method for a wide variety of field sizes and configurations. "Total" refers to methods that can be economically fitted to any size or shape of field; "partial" is for methods that can be fitted with difficulty or high expense, and "No" means that the method is only applicable for very large fields or of a certain geometry.
>
> Maintain by. This category gives some feeling for the complexity of the technologies in terms of overall physical sustainability. It does this by indicating who has the capacity to maintain, in a practical sense, the operability of the irrigation application equipment. *Farmer* indicates that the equipment could be easily maintained and fixed by an irrigator in the U.S., or

by a traditional farmer in less developed countries. *Grower* indicates that advanced skills normally associated with producers of high value crops are needed. *Shop* is used to indicate the need for merchants with some limited but special repair equipment. *Agency* indicates very sophisticated equipment that requires high level skills and expertise to maintain.

Table 2-1. Factors Affecting the Selection of Different Types of Modern Irrigation Systems for Use in Developing Countries (Modified from Keller and Bliesner, 1990).

| Method and Type | Divisibility | Maintain by | Risk | Mgmt & O&M Skill | Mgmt & O&M Effort | Ruggedness |
|---|---|---|---|---|---|---|
| **Surface** | | | | | | |
| Canal-Feed | | | | | | 0 |
| Basin | Total | Farmer/Grower | Low | Master | 5 | Lasting |
| Border | Total | Farmer/Grower | Low | Master | 6 | Lasting |
| Furrow | Total | Farmer/Grower | Low | Master | 10 | Lasting |
| Pump/Pipe-Feed | | | | | | |
| Basin | Total | Grower | Med | Master | 3 | Robust |
| Border | Total | Grower | Med | Master | 3 | Robust |
| Furrow | Total | Grower | Med | Master | 6 | Robust |
| **Sprinkler** | | | | | | |
| Periodic Move | | | | | | |
| Hand Move | Total | Grower | Med | Simple | 9 | Durable |
| End-Tow | Partial | Grower | Med | Medium | 5 | Durable |
| Side Roll/Wheel Line | Partial | Grower | Med | Medium | 6 | Durable |
| | No | Agency | High | Master | 5 | Fragile |
| Gun/boom | Partial | Shop | Med | Medium | 8 | Durable |
| Undertree | Partial | Farmer | Med | Simple | 9 | Durable |
| Fixed Position | | | | | | |
| Solid set Field Crops | | | | | | |
| Portable | Total | Grower | Med | Simple | 5 | Durable |
| Permanent | Total | Grower | Med | Medium | 1 | Durable |
| Orchard | | | | | | |
| Undertree | Total | Grower | Med/High | Medium | 2 | Durable |
| Overtree | Total | Grower | Med/High | Medium | 2 | Durable |
| Continuous Move | | | | | | |
| Traveling Gun | Partial | Agency | High | Master | 4 | Sturdy |
| Center Pivot | No | Agency | High | Complex | 1 | Sturdy |
| Linear Move | No | Agency | High | Complex | 2 | Sturdy |
| **Drip/Micro** | | | | | | |
| Orchard/Vineyard | | | | | | |
| Surface Drip | Total | Grower | High | Complex | 2 | Sturdy/Fragile |
| Microspray | Total | Grower | High | Complex | 2 | Sturdy/Fragile |
| Subsurface Drip | Total | Grower | High | Very Complex | 3 | Fragile |
| Row Crop | | | | | | |
| Surface Drip | Total | Grower | High | Complex | 3 | Sturdy |
| Subsurface Drip | Total | Grower | High | Very Complex | 4 | Fragile |

<u>Risk</u>. This category addresses the risk to crops due to equipment breakdown. A low risk indicates equipment that can be crudely patched and still work, as opposed to a high risk such as occurs with gearboxes on center pivots and

filters on drip systems. When gearboxes or filters fail, the complete crop in the field can be at risk for many days.

Management and O&M Skill. This category emphasizes management (rather than equipment) requirements to obtain reasonable irrigation efficiencies. It also accounts for the day-to-day maintenance that is required for long-term usage of the equipment. *Simple* indicates only elementary skills are needed, *Medium* indicates considerable skill, *Master* indicates that much hands-on field experience is necessary to spread the water or move the equipment, and *Sophisticated* indicates that sophisticated technical skills and reading abilities are necessary.

Management and O&M Effort. This indicates the relative management time and labor required to operate, manage, and maintain the irrigation system. A value of 1 indicates minimal effort; 10 indicates the highest level of effort.

Ruggedness. This indicates the durability of the equipment, as well as the sensitivity to breakdown and the level of spare parts and services needed. *Lasting* is used for gravity systems, since they never completely break down to the point of being completely inoperable. *Robust* systems have few mechanical parts. *Durable* systems rarely break down, but have some spare parts requirements. *Sturdy* systems require careful handling and maintenance and considerable spare parts. *Fragile* systems have delicate components that malfunction when handled incorrectly and require a considerable inventory of spare parts.

## **Physical Conditions**

The most obvious site conditions, and usually easiest to quantify, are the physical conditions. This information is needed for both selection and design. The importance in the selection process is to understand how the methods under consideration relate differentially to the physical conditions, identifying their impact on system design and performance. The physical conditions can be categorized in 5 areas: (1) Crops, (2) Water Supply, (3) Land, (4) Climate, and (5) Energy Supply. A checklist of the physical conditions to consider is presented in Table 2-2. Table 2-3 qualifies the conditions that are of particular concern for various irrigation methods and identifies which ones have a negative influence.

Crops and Cultural Practices. The crops being grown and the farming system used have a significant impact on system selection. The impact of the irrigation method on crop considerations such as crop height, germination, foliage disease and climate modification, and cultural practices including cultivation, herbicide and fertilizer application must be considered. Crop rotation is an extremely important consideration. Some irrigation methods may be suitable for some of the crops in the rotation, but not for all.

<u>Water Supply</u>. The source, quantity, quality, reliability and delivery schedule for the water supply must be carefully assessed in selecting irrigation methods. Water supplies that are not dependable preclude large investments in sophisticated irrigation equipment. However, supplies that are dependable but limited in quantity suggest the need for a highly efficient irrigation method.

Frequency, rate and duration of farm water delivery has an appreciable impact on the selection and design of an irrigation method, its costs, its labor requirements and its potential application efficiency. It may be necessary to make water delivery schedule changes to allow better irrigation, to permit use of alternate methods, or to make more economical use of labor and/or energy. A small continuous stream of water from a well or irrigation district is practical for a sprinkler system but not for a method needing large intermittent streams -- unless an on-farm reservoir is available (Jensen, 1981). A large flow rate at an infrequent interval is best utilized by surface methods. In general, rigid water delivery schedules inhibit efficient irrigation with any method.

The firmness of a water supply is an important issue because it impacts the economics and the crop selection. In areas with a variable irrigation water supply, farmers frequently leave fields fallow and/or adjust their crop selections to match the water supply.

<u>Water Quality</u>. Water quality is a strong consideration in the selection of irrigation methods. Adequate data on water quality are essential for the selection process. All irrigation water contains some dissolved solids and significant build up of these salts can occur without proper irrigation method selection and operation. The leaching capability of the irrigation method must be considered, and becomes increasingly important for higher water salinity in arid areas.

Specific ion toxicity can be a problem for some crops. For example, high levels of boron are damaging to many crop. Waters with high chloride levels should not be applied to the foliage of many fruit crops. A high ratio of sodium and magnesium to calcium can adversely affect soil structure and may reduce intake rate (intake rate) in many soils. Very pure irrigation water (EC < 0.2 dS/m) also reduces intake rates in many soils - causing difficulties with all irrigation methods.

Reuse of both tailwater (surface runoff) and tile water (subsurface drainage water) is increasing in many irrigation projects. These reuse waters are typically higher in silt and salt than the original supply water.

Physical contaminants such as suspended silt and debris, and organic particles such as algae and bacterial slimes can adversely affect some systems. Particulates may plug or abrade nozzles or orifices in sprinkler and drip/micro systems. Floating debris may cause trouble in other systems which discharge water through larger gates or openings. Different levels of filtration are required to prevent these problems.

Land. The texture, uniformity, depth, intake rate, erodibility, salinity and drainability of the soil must all be considered when selecting potential irrigation methods. The interaction between cultural practices and irrigation as related to these soil factors must be given due consideration. Some methods require additional soil information. For example, center pivot and linear move irrigation require information on the bearing strength of the soil under wet irrigation conditions to determine the ability of the soil to support the weight of the systems.

Field slope must be known to determine acceptable irrigation methods. Some degree of land leveling or smoothing may be required for some methods. Slope will also impact the hydraulics of piping systems and, for many systems, the uniformity of application. Irregular topography will have a similar affect, influencing field size and shape as well as leveling requirements for those methods requiring a uniform slope.

Field shape and physical obstructions may preclude the selection of some irrigation methods. The locations of buildings, overhead electrical lines and buried pipelines within fields to be irrigated and their boundaries are important to the selection process. Buried pipelines may limit the depth of cuts in land leveling. Overhead electrical lines will interfere with the operation of some mechanized sprinkler systems. Buildings and field boundaries will impact most irrigation methods.

If there is a potential for natural flooding in the area being considered for irrigation, the effects of potential floods on the methods being considered must be assessed. For example, if frequent floods with high flow velocities occur, then severe erosion may take place. In this event, expensive re-leveling would be needed for irrigation methods requiring uniform slopes. Excessive deposition of sediment may damage distribution and drainage facilities.

The water table depth should be known in any irrigation planning effort. Both the present and future water table depths must be known, requiring assessment of the irrigation method's potential impact on the water table. Highly efficient irrigation methods will cause less change in the water table. A moderately shallow water table allows consideration of water table management as an irrigation method, provided salt accumulation is not a problem.

Climate. Climatic data are essential to the process of irrigation method selection. Precipitation patterns determine the required degree of irrigation (full or supplemental) as well as the adaptability of some irrigation methods to these climatic patterns. Temperature, frost conditions, humidity and wind all impact the irrigation method selection process.

Irrigation modifies the climatic environment of the crop by augmenting natural precipitation. However, some irrigation methods have additional capabilities to perform freeze protection, bloom delay or to provide cooling or warming during critical temperature periods. If climatic modification is important, the capability of the various irrigation methods being compared should be considered.

Energy Availability and Reliability. Irrigation methods that require energy for pumping or maintaining control systems require a dependable energy source. If an energy source is not available, or is unreliable, these irrigation methods should not be considered. All reliable energy sources should be considered on the basis of cost, availability, and convenience.

Table 2-2. Physical Site Conditions to Consider in Irrigation System Selection.

| **CROPS** | **WATER SUPPLY** |
|---|---|
| Crops grown | Source |
| Crop rotation | Quantity |
| Crop height/root volume | Quality- salinity |
| Cultural practices | - sediments |
| Disease potential | - organics |
| Pests | Reliability |
| Water requirements | Delivery schedule- frequency |
| Climate modification | - rate |
|  | - duration |
| **LAND** | **CLIMATE** |
| Field shape | Precipitation |
| Obstructions | Temperature |
| Topography | Frost conditions |
| Soil    - texture | Humidity |
|   - uniformity | Wind |
|   - depth |  |
|   - intake rate |  |
|   - moisture |  |
|   - holding capacity | **ENERGY** |
|   - erodability | Availability |
|   - salinity | Reliability |
|   - drainability |  |
|   - bearing strength |  |
| Flood hazard |  |
| Water table |  |

Table 2-3. Suggested Guide for Selection of Irrigation Methods. (0) Indicates No Influence on Selection, (+) Indicates Possible Reasons for Preference, and (-) Indicates Possible Reason for Choosing Alternate Method.

|  | Surface | | | | | Drip/Micro | | | | Sprinkler | | | | | |
|---|---|---|---|---|---|---|---|---|---|---|---|---|---|---|---|
|  | Furrow | Modern furrow (a) | Border strip | Basin | Basin paddy | Bubbler low-head | Micro spray | Drip | Row Crop Drip | Linear move | Center pivot | Traveling gun | Side roll | Hand move | Solid set |
| **Physical conditions** | | | | | | | | | | | | | | | |
| Crop type | | | | | | | | | | | | | | | |
| corn | 0 | 0 | 0 | 0 | - | - | - | - | 0 | 0 | 0 | 0 | - | - | - |
| cotton-humid | 0 | 0 | 0 | 0 | - | - | - | - | - | 0 | 0 | - | - | - | 0 |
| cotton-arid | 0 | - | 0 | 0 | - | - | - | - | 0 | - | - | - | - | - | 0 |
| alfalfa | - | 0 | + | + | - | - | - | - | - | 0 | 0 | 0 | 0 | 0 | 0 |
| small grain | 0 | 0 | 0 | 0 | - | - | - | - | - | 0 | 0 | 0 | 0 | 0 | 0 |
| potatoes | - | 0 | 0 | 0 | - | - | - | - | 0 | + | + | - | 0 | 0 | + |
| rice | - | - | - | - | + | - | - | - | - | - | - | - | - | - | - |
| vegetables | 0 | 0 | 0 | 0 | - | - | - | - | + | 0 | 0 | - | 0 | 0 | 0 |
| other row | 0 | 0 | 0 | 0 | - | - | - | - | 0 | 0 | 0 | 0 | 0 | 0 | 0 |
| orchards/vineyards | 0 | - | - | + | - | + | + | + | - | + | + | 0 | - | 0 | + |
| rotation | 0 | 0 | 0 | 0 | - | - | + | - | 0 | + | + | - | 0 | 0 | 0 |
| climate modification | - | - | - | - | - | - | - | - | - | - | - | - | - | - | + |
| cultural/machinery operations (equip. In the way, field length, etc.) | 0 | 0 | 0 | - | - | 0 | 0 | 0 | + | - | - | - | - | - | 0 |

(a) Mechanized furrow irrigation refers to irrigation with a cutback stream, surge flow, cablegation, etc.

Table 2-3. *(Continued)*

| Land | Surface | | | | | Drip/Micro | | | | Sprinkler | | | | | |
|---|---|---|---|---|---|---|---|---|---|---|---|---|---|---|---|
| | Furrow | Modern furrow (a) | Border strip | Basin | Basin paddy | Bubbler low-head | Micro spray | Drip | Row Crop Drip | Linear move | Center pivot | Traveling gun | Side roll | Hand move | Solid set |
| odd-shaped fields | 0 | 0 | 0 | + | + | + | + | + | + | − | − | + | − | 0 | 0 |
| obstructions in field | 0 | 0 | 0 | 0 | 0 | 0 | 0 | 0 | 0 | − | − | + | − | 0 | 0 |
| high water table | − | − | − | − | − | + | + | + | + | 0 | 0 | 0 | 0 | 0 | 0 |
| undulating slope with shallow soils | 0 | 0 | 0 | − | − | + | + | + | + | + | + | 0 | 0 | + | + |
| steep slopes | 0 | 0 | 0 | − | − | − | − | + | + | 0 | 0 | 0 | 0 | + | + |
| steep, rocky slopes soil | − | − | 0 | − | − | − | + | + | + | + | + | − | − | + | + |
| sandy, high infiltration soils | 0 | 0 | 0 | 0 | + | + | 0 | 0 | 0 | 0 | 0 | + | + | 0 | + |
| loamy moderate infiltration soils | 0 | 0 | 0 | 0 | 0 | 0 | − | 0 | + | − | − | − | 0 | 0 | 0 |
| clay, low infiltration soils | 0 | 0 | 0 | − | 0 | 0 | + | + | + | + | + | + | + | + | + |
| highly non-uniform soils | − | − | − | 0 | 0 | + | + | + | + | + | + | + | + | + | + |
| low water-holding capacity soils | 0 | 0 | 0 | 0 | 0 | 0 | + | + | + | 0 | 0 | 0 | + | + | + |
| saline soil | − | − | − | − | 0 | + | + | + | + | 0 | 0 | 0 | − | − | 0 |
| poorly drained soil | 0 | 0 | 0 | 0 | 0 | + | + | + | + | 0 | 0 | − | 0 | 0 | 0 |
| highly erodible soil | 0 | 0 | 0 | 0 | 0 | + | + | + | + | 0 | 0 | − | 0 | 0 | 0 |
| low-bearing capacity soil | 0 | 0 | 0 | 0 | 0 | 0 | 0 | 0 | 0 | − | − | − | 0 | 0 | 0 |

Table 2-3. *(Continued)*

| | Surface | | | | | Drip/Micro | | | | Sprinkler | | | | | |
|---|---|---|---|---|---|---|---|---|---|---|---|---|---|---|---|
| | Furrow | Modern furrow (a) | Border strip | Basin | Basin paddy | Bubbler low-head | Micro spray | Drip | Row Crop Drip | Linear move | Center pivot | Traveling gun | Side roll | Hand move | Solid set |
| ***Water supply*** | | | | | | | | | | | | | | | |
| groundwater well | O | O | O | O | . | + | + | + | + | + | + | + | + | + | + |
| delivered surface water - flexible | O | O | O | O | O | O | O | O | O | - | - | - | - | - | - |
| delivered surface water - rigid | O | O | O | O | O | - | - | - | - | - | - | - | - | - | - |
| delivered surface water - continuous flow rate | - | - | - | - | O | + | + | + | + | + | + | + | + | + | + |
| unreliable rate and timing | O | O | O | O | - | O | - | - | - | - | - | - | - | - | - |
| high sediment load | O | O | O | O | O | - | - | - | - | - | - | - | - | - | - |
| high organic matter | O | O | O | O | O | O | O | - | - | O | O | O | O | O | O |
| high salinity | O | O | O | O | O | O | O | + | O | - | - | - | - | - | - |
| wastewater | O | O | O | O | O | O | O | - | - | O | O | O | - | - | O |
| large rate of flow | O | O | O | + | O | O | O | O | O | - | - | - | - | - | - |
| small rate of flow | O | O | O | - | O | + | + | + | + | + | + | + | + | + | + |
| ***Climate*** | | | | | | | | | | | | | | | |
| high rainfall | O | O | O | - | O | O | O | O | O | + | + | + | + | + | + |
| low rainfall | O | O | O | O | O | O | O | O | O | - | - | - | - | - | - |
| high temperature - humid | O | O | O | O | O | O | O | O | O | + | + | + | + | + | + |
| high temperature - arid | O | O | O | O | O | O | O | O | O | - | - | - | - | - | - |
| windy | O | O | O | O | O | O | O | O | O | - | - | - | O | O | - |
| frost conditions | - | - | - | O | - | O | + | O | O | O | O | O | O | O | + |

Table 2-3. (Continued)

| | Surface | | | | | Drip/Micro | | | | Sprinkler | | | | | |
|---|---|---|---|---|---|---|---|---|---|---|---|---|---|---|---|
| | Furrow | Modern furrow (a) | Border strip | Basin | Basin paddy | Bubbler low-head | Micro spray | Drip | Row Crop Drip | Linear move | Center pivot | Traveling gun | Side roll | Hand move | Solid set |
| ***Energy*** | | | | | | | | | | | | | | | |
| scarce or unreliable | o | o | o | o | o | o | – | – | – | – | – | – | – | – | – |
| ***Social/institutional conditions*** | | | | | | | | | | | | | | | |
| low labor skills | o | – | o | o | o | o | – | – | – | – | – | – | – | o | o |
| low parts availability | o | – | o | o | o | – | – | – | – | – | – | – | – | – | – |
| little technical assistance available | o | – | o | o | o | o | – | – | – | – | – | – | – | o | o |
| low labor availability | – | – | – | + | + | + | o | o | o | + | + | o | – | – | + |
| automation potential | + | o | + | o | o | o | + | + | + | + | + | o | o | o | + |
| vandalism potential | o | o | o | + | + | o | – | – | – | – | – | – | – | o | o |
| low management skills | – | – | – | + | + | o | – | – | – | – | – | – | o | o | o |
| environmental concerns | – | – | – | – | – | o | o | o | o | – | o | o | o | o | o |
| land transformation | – | – | – | – | – | o | o | o | o | – | o | o | o | o | o |
| chemical use | – | – | – | o | o | + | + | + | + | o | o | o | o | o | o |
| sustainability | o | o | o | + | + | o | o | o | – | – | – | – | o | o | o |

## Economic Considerations

As discussed earlier, economic efficiency is central to the irrigation system selection process. While other considerations are important, most, if not all, have economic implications. Regardless of the importance of non-economic goals of an irrigation development, some economic analysis will be required.

The economic analysis process will be discussed in detail in a later section of this chapter. The economic data discussed here are required to define the conditions under which the economic analyses will be made. The economic data required falls into two categories: site dependent and system dependent.

Site dependent economic parameters are those that are not influenced by the system development, but which are necessary to determine the relative economics of development. They include: interest rate (real and nominal), labor cost, energy cost, energy inflation factor, general inflation factor, property tax (on irrigation equipment), water cost, land value and return to irrigation for each crop.

Interest rates are often categorized as "real" or "nominal". Nominal rates are the current rates of interest charged by the lending institution that will be the source of money. The rate includes an inflationary component and a risk, management and profit component. The real rate is inflation free and, therefore is less than the nominal rate by the long-term inflation rate. The real rate is usually in the range of 3 to 7%. The real rate is used to determine the annualized cost of capital expenditures that tend to appreciate, such as land values and permanent improvements to the land, such as land leveling. The nominal rate is used to determine the annualized cost of capital expenditures that tend to depreciate or reach technical obsolescence with little or no salvage value at the end of the project life.

Land value must be considered when comparing one system that does not irrigate all the land to another irrigation system that irrigates all the land (e.g., a center pivot that does not irrigate corners vs. a drip irrigation system that irrigates the whole field). Crop returns to irrigation are the net returns after all production and land costs associated with irrigation are deducted. These returns are needed to compare the relative economics of systems with different uniformities that may have an impact on yield.

The energy inflation factor is important in balancing capital cost with operating cost. Inflation factors should also be included for other irrigation input costs, such as labor and water cost. These inflation factors are used in the same manner as an energy inflation factor. The long-term general inflation rate is usually adequate to approximate these values.

System dependent parameters are those that relate directly to the system and are identified individually for each system. They include: system components costs, system components lives, labor requirement, energy requirement and maintenance cost. The physical life of some components may be longer than the expected technical obsolescence of the irrigation method. In these cases, it is practical to use an expected economic life rather than the full physical life. For example, PVC pipe has a very long (50-100 years) life, but the system may reach obsolescence in a much shorter time, so the shorter time should be used in computations.

Labor and energy requirements are not economic parameters directly, but determine the economics of system operation and are listed here for that reason.

## SYSTEM PRE-SELECTION

Once the goals are defined and the site data determined, pre-selection or screening of systems can proceed. The detailed selection process is too rigorous to complete for every system, nor is it necessary for systems that obviously can not fit either the goals or the site conditions. As experience is gained for a particular location, it becomes easier to eliminate systems during the pre-screening, since more is known about the likelihood of a particular system or configuration fitting the site conditions. However, care should be taken to avoid elimination of systems that may be likely candidates but were eliminated in a previous analysis, when the conditions are not sufficiently similar to allow such extrapolation of results.

The process is one of matching the capabilities of the potential irrigation systems to the goals of development and site conditions. The various strengths and limitations listed in the following sections can be weighed against the site conditions identified to establish suitability. Any systems that would obviously fail to meet the goals can be eliminated. Table 2-3 lists the relative strengths and limitations of systems to given site conditions and may be used as a guide in absence of local experience with particular methods.

## FEASIBILITY DESIGN AND ECONOMIC ANALYSIS

Once the pre-selection process is complete, detailed analyses of the remaining selected systems are performed. If a large number of systems are pre-selected, this may be a two-step analysis with less detailed analysis, especially of the less promising systems, as the first step. If the number of systems selected for analysis is not large, then each system should be designed and the economic conditions determined.

Once designed, each system will require a full economic analysis, within the constraints of the goals selected, for comparison and final selection. This economic analysis is most easily completed on an annualized cost basis. The costs and returns resulting from each system should be determined and compared as shown in Figure 2-1.

The first step is to complete the designs and determine the total annualized costs of the systems. The second step is to compute the expected returns to the development.

The annualized capital costs are computed using the capital recovery factor (CRF) described in equation 2-1 for the life of the system component and the nominal interest rate. Each component will have a capital recovery factor associated with its life.

$$CRF = \frac{i(1+i)^n}{(1+i)^n - 1} \qquad (2-1)$$

Where   $i$ = annual interest rate
$n$ = the economic life of the component

Multiplying the total project component cost by the CRF value will give the annual cost of the components.

If there are capital items, including land cost, that do not depreciate, use the inflation free or "real" interest rate to account for the fact that they will hold their value.

Energy costs should be adjusted for estimated inflation by using the EAE, the equivalent annualized cost factor of escalating energy taking into account the time value of money over the life cycle (Keller and Bliesner, 1990).

$$EAE = \left[\frac{i}{(1+i)^n - 1}\right] \cdot \left[\frac{(1+e)^n - (1+i)^n}{(1+e) - (1+i)}\right] \qquad (2-2)$$

Where   $i$ = annual interest rate
$n$ = the economic life of the component
$e$ = decimal equivalent annual rate of energy escalation

The equivalent annual cost of energy equals the EAE times the current cost of energy.

The expected average annual maintenance costs over the life of the components should be estimated and included in the analysis. Table 2-4 lists some typical maintenance costs expressed as a percentage of the original capital costs for major system components. Use these values if local costs are not available.

The labor cost to operate each system type must also be computed. Typical labor requirements for various irrigation methods are given in Table 2-5. These values are expressed as man-hours per irrigation per hectare for in-season costs. Since there is cost associated with starting the systems at the beginning of the season and storing the systems at the end of the season, a pre- and post-season requirement is also shown. The operating cost of some systems is not easily computed on a per-irrigation basis. Refer to the footnotes under the table for explanation of the exceptions. The labor shown is for operation only and not maintenance. Maintenance labor is covered in the maintenance costs shown. Many methods have significant variability in labor requirements depending on local conditions. Local experience should be used to refine the values in Table 2-5.

Labor costs are computed using a local cost for labor and the man-hours shown. For comparing systems with different labor costs, the expected inflation in labor cost should be used. The equivalent annual labor cost over the life of the project may be computed by using EAE computed with the expected labor inflation rate, rather than the energy inflation rate. This factor could be termed equivalent annual labor cost (EAL). Care should be taken to use long term inflation rates to avoid bias in the analysis due to short-term phenomenon.

The other annual irrigation costs would include taxes on irrigation equipment, if any, and water costs. If these costs are inflating, they should be adjusted for inflation for comparison purposes as discussed above, only if it is anticipated that they will be different among the methods analyzed.

The returns to the project are computed for the crops to be grown. If the systems being compared have markedly different application uniformities, then the yield impact of these uniformity differences should be determined if crop production functions are available. In this case, the gross returns and production costs for each crop will be required. If the uniformities are not markedly different, then only the net returns from the crops are required.

The net returns are computed by subtracting the average annual costs from the average annual benefits. If the economic goal is to maximize net return, then the system with the largest net return best meets the goal.

The benefit/cost (B/C) ratio is computed by dividing the annual benefits by the annual costs. If the goal is to maximize return on investment, then the system that yields the largest B/C ratio best meets the goal. It is possible, even likely, to have one system yield the best net returns and another have the highest B/C ratio.

Table 2-4. Typical Economic Lives and Maintenance Costs for Irrigation System Components.

| Component | Economic Life (yrs) | Maintenance, (% of Cost) |
|---|---|---|
| **Surface Irrigation** | | |
| Buried pipe | 30 | 1 |
| Gated pipe, aluminum | 10-20 | 3 |
| Gated pipe, PVC | 5-10 | 5 |
| **Sprinkler Irrigation** | | |
| Lateral | | |
|   Hand move | 15 | 2 |
|   End-tow | 10 | 3 |
|   Side roll | 15 | 2 |
| | 15 | 4 |
| Hose-fed | 5/20 | 3 |
| Traveling gun | 10 | 6 |
| Center pivot | | |
|   Standard | 15 | 5 |
|   w/corner | 15 | 6 |
| Linear move | 15 | 6 |
| Solid set | | |
|   Portable | 15 | 2 |
|   Permanent | 20 | 1 |
| **Drip/Microirrigation** | | |
| Orchard | | |
|   Drip or Microspray | 15/25 | 5 |
| Row-crop | | |
|   Multiple year emitters or tape | 6/15 | 6 |
|   Disposable tape | 1/(3-15) | 10 |
| **Other Components** | | |
| Buried PVC mainline | 20-40 | 1 |
| Steel mainline | 10-20 | 1 |
| Aluminum mainline | 10-20 | 2 |
| Electric pumps | 15 | 3 |
| Diesel/gas pumps | 10 | 6 |
| Wells | 25 | 1 |

Notes:
Where two lives are shown with a slash, the first number is for above ground components and the second for below ground components. These values are approximate. Local experience and local operating conditions should be considered when available.

Table 2-5. Average Operating Labor Requirement for Sprinkler and Drip/Micro Irrigation Systems.

| System | Pre & Post Season[1] man-hrs/ha | Per Irrigation man-hrs/ha |
|---|---|---|
| **Sprinkler Irrigation** | | |
| Center pivot | .12 | .05[2] |
| Linear move | | |
|   Ditch fed | .12 | .10 |
|   Hose fed | .15 | .15 |
|   Pipe fed | .12 | .07 |
| Side move | .49 | .62 |
| Side roll | .25 | .86 |
| Traveler | .25 | .62 |
| Hand move | .25 | 1.73 |
| Solid set | | |
|   Portable | 2.47[3] | .15 |
|   Permanent | .25 | .15 |
| **Microirrigation** | | |
| Orchard | | |
|   Drip | .25 | .05[4] |
|   Micro | .25 | .05 |
| Row crop | | |
|   Permanent | .50 | .05[4] |
|   Disposible | 2.00[5] | .05[4] |

[1] The amount shown for each pre or post season operation.
[2] Assumes 1" net application or greater.
[3] Add 2.47 hrs for each mid-season move.
[4] Computed using 1 hr/day for each 60 ha & 2 day irrigation interval.
[5] Assumes tubing is laid during planting by machine.

## FINAL SELECTION

Final selection requires identification of the system(s) that best meets the development goals. Since pre-screening eliminates the systems that do not meet the non-economic goals, final selection usually reduces to selection of the system that either returns the greatest net benefits, or provides the best return on investment, depending upon the goal selected.

In some cases, there may be a clear choice. However, in many cases, more than one system may meet the goals equally well, given the uncertainties in the analysis. In these cases, evaluation of the components of the design will be necessary. An assessment of the risks associated with the operational conditions of the methods in terms of future operation would add information that would be relevant to the final selection.

In all cases, the final selection should not be made by the designer alone. The comparative data for the better choices should be presented to the owner and/or operator and the decision made jointly.

In making the final selection, variations on the systems that appear to be the best may also be necessary. In the example systems, splitting the pumping stations and pipelines between the two sub-mains would have yielded lower energy costs, but would have increased capital costs. Adding corner systems to the center pivots, or examining alternate means of irrigating the corners would be other logical options to explore. By using the process outlined, many alternatives can be explored and compared. With these tools available, the process of design and selection can be unified to produce informed selection of systems to best meet development goals.

# CHAPTER 3
# SURFACE IRRIGATION

## DESCRIPTION

Surface irrigation refers to a large group of irrigation methods in which water is distributed by gravity over the surface of the field. Water is typically introduced at the high point or along the edge of a field and allowed to cover the field by overland flow. The efficiency and uniformity of irrigation is dependent upon soil uniformity, quality of land grading, field topography and the control of the relationship between stream size, soil infiltration rate and duration of application.

The defining feature of surface irrigation methods is that the soil is used as the transportation medium (as opposed to pipelines or through the air, as with sprinklers). The soil also controls the depth infiltrated over time (as opposed to the application rate being controlled by sprinklers or emitters). Furthermore, the infiltration and advance characteristics of surface irrigated fields change with time, making it impossible to pre-determine many management recommendations. Irrigation control by field management is more important with surface methods than for the mechanical systems where design and equipment replace much of the need for intensive management.

Surface irrigation methods have two basic categories: ponded and moving water. The moving water methods require some runoff or ponding to ensure adequate infiltration at the lower end of the field. Tailwater return flow systems are often required by law to prevent runoff from farms. They also provide valuable tools for labor reduction and improved uniformity if designed properly. Table 3-1 lists the surface irrigation methods described in this chapter.

## TYPES OF SURFACE IRRIGATION METHODS

### Basin

Basin irrigation is a ponded water irrigation method used to apply water to a level area of land bounded by dikes (Figure 3-1). The soil surface is not kept flooded. Because the water is ponded until it infiltrates, there is no runoff. In rainfall areas consideration must be given to provide for draining unneeded water. Basin irrigation is used under a wide variety of names, including check flooding, level borders, check irrigation, check-basin irrigation, dead-level irrigation, and level-basin irrigation. This method of irrigation can be used for both field and row crops, often interchangeably, with or without interior ridges or with wider flat beds. It is also used on trees and vines. The soil intake rate should be the same within any one basin as infiltration uniformity is very sensitive to variation of intake rate. The basins need not be rectangular nor straight sided and the dikes need not be permanent. Under good management, a pre-determined volume of water is rapidly discharged into the basin.

Table 3-1.    Surface Irrigation Methods and Variations.

| Method | Variations |
|---|---|
| Basin | Flat-Planted Basins |
|  | Bedded or Furrowed Basins |
|  | Fill and Drain |
|  |  |
| Border Strip | Sloping Strips with Runoff |
|  | Low-Gradient and Blocked End Strips |
|  | Contour Strips |
|  | Contour Ditches (Wild Flood) |
|  | Guided |
|  |  |
| Continuous Flood and Ponding |  |
|  |  |
| Furrow | Traditional Sloping Furrows |
|  | Modern (Mechanized) Sloping Furrows |
|  | Level and Low-Gradient Furrows |
|  | Contour Furrows |
|  |  |
| Corrugations |  |

Figure 3-1. Large Scale Basin Irrigation with Corner Inlet.

The method variations consist of flat-planted and bedded (ridged, channeled, furrowed) basins. The crops that have been successfully grown with level basins are nearly unlimited but soil related. Bedded basins are typically used for row crops where it is necessary to avoid inundation or when light applications are desired with bed widths being critical. Narrow ridges to wide beds are often used on which to plant vegetables, melons, cotton, corn, potatoes, sugar beets, and many other row crops.

The flat-planted basins are best suited to field crops and row crops that are not sensitive to inundation for short periods of time. Field crops such as alfalfa, wheat, sorghum, barley, cotton, etc., are often irrigated this way. This flat planting can help to eliminate salinity problems and facilitate heavy water applications. Orchard crops and vineyards can be planted flat and can also be planted on beds or islands.

Bedded basin irrigation is particularly suitable for row crops that require control of moisture within the beds. For example, level beds can facilitate uniform wetting and germination which may be very difficult with furrows but which is relatively easy to obtain in small basins. However, if the basins are very large, the large flow rates will fill the upstream ends of the channels very high with water during the advance phase, and may inundate the crop or seedbed.

Crop damage due to inundation is reduced by use of precisely leveled basins. Rain water and excess irrigation is uniformly distributed over the entire area rather than being concentrated in low spots or at the low end of the field, where crop damage typically occurs. However, since there is no runoff, excessive over application by irrigation or excess rainfall can cause crop damage. In areas with high intensity rainfalls and low intake rate soils, surface drainage systems must be considered.

Basins may have some advantages over other manually controlled irrigation methods because of ease of operation, simple equipment requirements, potentially low labor requirements and the capability to utilize large fixed rate streams. The method is most efficient with uniform soils, precise leveling, and large stream sizes relative to basin area. None of these are absolute requirements, but they tend to make basins more effective.

The irrigator or supervisor needs to know the incoming flow rate, the basin area, the desired application, and attainable uniformity in order to determine the irrigation set time accurately; a six minute error in a 60 minute set is a 10% error. The uniformity is related to the stream size, advance ratio and irregularity of the soil surface, and intake rate. Advance ratio (AR) is defined as:

$$AR = \frac{\text{time of advance}}{\text{time of irrigation}} \qquad (3\text{-}1)$$

The "time of irrigation" is typically the desired opportunity time for infiltration of the SMD. In practice, it is defined as the smallest opportunity time in the basin or furrow. "Opportunity time" is the time the water is in contact with the soil surface at a point in the field.

The irrigation labor requirements are potentially low where large streams are used intermittently. However, the intermittent labor must be usable at some other location or manner. The physical work involved generally amounts to opening a gate or port without flow adjustment. For a well designed system, little skill or knowledge by the irrigator is required, but a supervisor must clearly understand the relationships between flow rate, duration, and soil moisture deficit.

With level basin systems, particularly on row crops with ridge or beds, the channels may be open and interconnected on both ends by secondary ditches. Water moving in the faster advancing streams is collected in this secondary ditch and then flows back in the channels where the advance is slower. This reduces non-uniformity of inundation time. Water can also be supplied from both ends of a level basin.

Water can be distributed to several basins from a common corner. This can provide for easier operation and can facilitate water spreading to both sides. It may also reduce the length of farm ditches needed. In flat terrain, basins are usually set up on a regular grid pattern corresponding to property boundaries and soil texture changes. On sloping ground, basins are generally set up with the basin length parallel to the contours and the water supply running down the main slope. If the water supply canal is below the field surface and is checked up to deliver water, the same canal can be used to irrigate the basin and can also serve as a surface drainage canal.

When ridges or beds are used in a basin, secondary head ditches distribute water to individual channels by gravity. It is not necessary to divide the flow entering the basin, and the irrigation stream can enter the field through a single inlet. No irrigator intervention is required if the cultural operations are done properly.

Some means of erosion control at the inlet to the basin is required if the stream is large. However, standard erosion control structures are available. When the supply canal is below the field level, no erosion will occur since, to irrigate, the water level in the canal is simply checked up until water begins to flow into the channels or basin.

The water may be applied rapidly so that the basin is covered quickly for high uniformity. If the time required to cover the basin with water is short compared to the inundation time (small advance ratio 1:4 to 1:2), then the basin should have good infiltration uniformity within the accuracy limits of the land grading and soil uniformity. Any slope on the basin will cause water to stand in the lower area of the basin for a somewhat longer time. Inadequate leveling or leaving some slope may reduce construction cost but will generally lower the uniformity of water application. The small ± 1 cm undulation, even with laser controlled grading, can cause non-uniform infiltrated depths of the corresponding ponding amount.

Cultural practices can be performed in any direction within a level basin. A single irrigation ditch can supply basins on either side, thus reducing the length of ditches required. When buried pipelines are used, cultural practices are not restrained by basin length.

The construction of large basins with large cuts and fills is generally unnecessary and seldom economical. Narrow basins can be laid out on the contour even on relatively steep land, but at an increased cost and greater area out of production for high benches. Leveling requirements are largely a function of basin size and topography. Generally, the smaller the basin the less the grading requirement but the more difficult it is to utilize laser-controlled land grading equipment.

Basin sizes range from a few square meters (Figure 3-2) to 10 - 15 ha (25 - 40 ac). Basin size and length are often limited by topography, soil texture changes, infiltration, depth of topsoil (when extensive leveling is involved), available stream size, degree of land leveling, and farmer equipment capabilities. From a design standpoint, basin length is limited primarily by soil uniformity, intake rate, and available stream size as well as by potential for erosion. Large

Figure 3-2.   Small Scale, Hand Constructed Bedded Basin Irrigation.

basins must consider earth curvature in land grading, as a horizontal plane does not follow the earth's curvature.

In general, basin systems are designed for a given minimum depth of application. As the advance ratio becomes large, it becomes less efficient to apply small quantities of water. In some instances, the full basin may not be covered by a small water application. Typical irrigation depths for field crops are 70 - 100 mm (2.5 - 4.0 in) of water. Smaller applications may require shorter basins to maintain a satisfactory advance ratio. Larger applications give more uniform water distribution but may result in excessive deep percolation or excessive inundation times. Small application depths of 30 - 50 mm (1 - 2 in) are possible on some soils with bedded basins using narrow channels and wider beds or by using the "ponding" technique described in the Ponding Method Section.

Basin lengths should be designed according to soil intake rates and available stream size. Crops grown, water application depth, desired uniformity, stream size, and other practical considerations may further limit basin size and length. For coarse-textured, high intake rate soils, basin lengths should be limited to about 100 m (330 ft). For medium to fine textured soils, basin lengths of 400 m (1,320 ft) are often used. For soils with very slow final intake rates, surface drainage for excess application is required. The ponding method is preferable for these soils. Short basins facilitate lower advance ratios (between 0.25 and 0.35), which give better water distribution uniformities.

The stream size available will determine basin width necessary to provide an acceptable unit stream size. The unit stream size will affect the advance ratio and uniformity as they relate to length.

## Border Strip

Border strip (contoured, graded, and guided) irrigation consists of a sloping strip of land essentially level across the strip and bounded by borders (dikes, levees, ridges, etc.) to prevent the lateral spread of water (Figure 3-3). The borders are usually parallel for convenience of cultivation, but this is not necessary for irrigation, particularly for contour border strips. A relatively large stream of water is turned in at the upper end, spreads across the strip, and advances down the strip. When all pertinent items (unit stream size, soil moisture deficiency, intake rate, flow retardance, and length) are in proper relations, the stream is cut off when the advancing water is between about 0.6 of the length for slow intake soils to 0.9 of the length for fast intake soils if little or no tailwater is desired. The lower end of the strip is irrigated by water supplied from the temporary surface storage on the upper end. It is possible to have only a small amount of water run off (perhaps 10%) and the upper and lower ends are generally under irrigated relative to the middle portion. This scenario requires one specific combination of soil moisture depletion and flow rate.

There are several variations of the Border Strip method. These are briefly described below.

Contour border strips. Contour border strips go across the slope with a slight downslope longitudinal gradient. They are usually narrow to reduce land grading and form small benches at each border. They are typically not straight, but follow the "contour".

Graded border strips. Graded border strips are graded to a uniform slope lengthwise and usually level at right angles to the flow direction. Graded border strips involve more earth moving than other variations of border strips. Small variations in longitudinal and lateral slope are not greatly detrimental. They are the most common "modern" form of border strips, as the land grading is possible with laser guided equipment.

Guided border strips. Guided border strips requiring less land grading are applied on shallow soils and non-uniform topography. They are run nearly straight down slope following the topography with only small cuts and fills and are usually narrow strips to reduce or eliminate cross slope. Lengths are often compromised so typically they are managed more like furrows than border strips with their unique recession curves.

Figure 3-3.  Border Strip Irrigation with Pulled-up Borders. (Blocks Are Necessary to Keep Water from Channeling Down Furrows.)

All variations of border strip irrigation flood the soil surface and will cause some soils to form a crust. Such a crust may inhibit sprouting of seeds. For some crops, wetting of the plant may be detrimental due to fungi and diseases that may find a suitable moist environment. This tends to be more of a problem on the finer texture soils that remain moist for a long period.

All crops that are not limited by the above cultural considerations can be adapted to the border strip method. Field crops planted in rows: grain, hay, pasture, crops planted flat: orchards and vineyards sometimes planted on the border or in the middle or edge of the border strip, are easily irrigated this way. Ponding at the lower end of a blocked-end border strip frequently is detrimental to crops, especially if ponding continues for longer than a few hours duration, as may occur on fine textured soils.

Annual crops with changing root zones and allowable soil moisture deficiencies can be grown with the border strip method only by compromising on application efficiency or using a supplemental supply lines because of the great change in SMD and duration of irrigation during the season. Border strips are best suited for pasture, alfalfa, orchards, vineyards, and other crops that have constant MAD requirements.

Border strip irrigation is the most complicated of all irrigation methods. While the method has a very high potential application efficiency, high efficiency is seldom obtained. This is due to inadequate knowledge of the need for practical compromises of the management allowed soil moisture deficit (MAD) (design control) to the actual soil moisture deficit (SMD) on the date of irrigation, and the optimum stream size and distance and time to cutoff. Crop retardance and the intake rate vary during the season though the final intake rate and recession curve do not. Length, width, duration, and flow rate are management controllable items. High Distribution Uniformity (DU) is almost always attainable by using a stream size to make the advance curve essentially parallel to the unique controlling recession curve of each strip, but attention to detail is necessary to accomplish this. Since the intake rate changes during the season, different durations will be needed.

A large stream size allows a large area covering many strips to be set at one time, increasing labor efficiency. It will require a larger distribution system with a higher capital cost. The labor needed for a conventional small flow border strip system using siphons from a ditch is appreciable. Gates from the ditch greatly reduce the work but the irrigator still has to go to the field for each set. If a large stream is available to allow large sets and the water control is easily accomplished because of adequately designed controls, very little labor is required. Piped lines may be desirable and economical. It is practical for one man to control irrigation on 250 to 400 ha (600 to 1,000 ac), opening and closing a few gates or valves. If a regulating reservoir is incorporated, the time and duration of application can be at the convenience of the irrigator and large streams are readily available. Several fields can be set in sequences to run simultaneously when large, flexible streams are available

Undulating topography and shallow soils do not respond well to grading to a plane. For this condition, guided border strips downhill or contour border strips (benches) across the slope may be practical. The cross slope procedure creates a hazard of the borders breaking, especially in soil that cracks upon drying or with sandy erodible soils. The downhill procedure may be subject to erosion from irrigation or rainfall unless a suitable crop cover is established or strip lengths are short.

Border strips may have the surface slope gradient on the lower end reduced or even made level, or the ends may be blocked to reduce or eliminate runoff. However it may often only be converted to deep percolation loss. Care must be taken to assure that water is not ponded for a detrimentally long time.

Land grading to a plane surface with slope in two directions of different values may be desirable when row crops and alfalfa or pasture are rotated. Close growing or solid crops can be irrigated in the direction of the steeper gradient, and the row crops with furrows can be irrigated in the direction of the milder gradient. Such a design requires water delivery to two sides of a field, which increases the cost.

Grading requirements for border strip irrigation are less than for basins and are dependent upon field size and design. Allowing cross slope reduces the volume of earth moved, as does the use of guided border-strips. Having gradients closely approximating topographical slopes is more economical than forcing surface slopes to preset gradients. There is negligible difference in capabilities between gradients of .002 and .003. Moreover, narrow contour benches or strips require less earth moving than wider ones having higher benches.

Variable longitudinal or cross slope is not greatly detrimental if variations are moderate (.002 to .003) as both the advance and recession tend to remain parallel and land preparation costs may be reduced. For non-uniform soils within a desired field size, over-cutting and replacement of topsoil to obtain more uniform surface conditions will increase efficiency. However, the cost may make this option less economical than considering other irrigation methods if the volume of earthwork is large.

The difficulty of designing the length of the strips to conform to the limitations of intake rate, stream size, management allowed soil moisture deficit (MAD), flow retardance, and slope is compounded by the presence of fixed field boundaries. Appreciable variation in length can be accommodated if the management allowed deficit (MAD) can be varied or a lower efficiency accepted. Short lengths can be used for light irrigations and long ones for heavy application, but both cannot be done at high efficiency on the same strip. A supplemental supply at half length may effectively shorten a long strip. If very short strips are physically required, it may be better to use either the basin or furrow irrigation methods. Otherwise, a great deal of runoff may occur.

For rapid intake rate soils, strips may be only 100 m (330 ft) long. For slow intake conditions, strip length may need to be 800 to 1,000 m (2,600 to 3,300 ft) to provide a flow duration that allows time for an adequate irrigation to infiltrate. Also, stream sizes need to be varied. Lengths and stream sizes other than the ideal are practical but at reduced efficiency. Changing the slope or direction may be practical. Uniform intake rate and water holding capacity in the root zone are very desirable within any one unit that requires similar management, especially in the downslope direction.

With cracking soils where the cracks must be filled before the flow will advance, a depth slightly more than that needed to fill the cracks is the minimum that can be applied. Changing the soil moisture deficiency and crack size at the time of irrigation has great impact. After the cracks have been filled, the intake rate typically becomes very slow. Little water, possibly not even half again as much, can be infiltrated and the strip can be as long as 1,000 m (3,300 ft) and advance and recession need not closely parallel each other.

Border dike construction may be done seasonally or they may be permanently installed. Their spacing and cost is related to cultural practices such as crop spacing, equipment width, and the available stream size.

**Continuous Flood (Basin Paddy)**

Continuous flood irrigation is an irrigation method by which water is continuously ponded, although fields may be supplied water either intermittently or continuously. Fields consist of level or nearly level areas bounded by dikes to retain the ponded water and are often benched on slopes. Continuous flood irrigation is used specifically for rice that requires or can adapt to saturated, inundated soils. The crop is typically flat planted.

In many areas continuous flood systems are irrigated with water cascading through the upper fields to make up for water lost by seepage and evapotranspiration in the lower fields. Better water control is possible when a separate supply ditch to each field is used. This is also important during the dry season when a variety of crops may be grown in different fields and the water requirements for each field are different. The water delivery system should be capable of supplying and controlling variable flow rates as different fields are ready for flooding on different dates. High flow rates at this time are important since the intake rates are very high during initial flooding.

Continuous flood is best applied on soils with low intake rates. Seepage is often limited by a restricting layer (or plow pan) which develops over years of cultural practices with saturated conditions. Soil uniformity is generally not a problem once a plow pan has developed or a water table has been established.

Water for continuously flooded fields is typically applied at slow, continuous rates, sufficient to keep up with water requirements -- after the initial field flooding. In some areas, water is supplied periodically. This has little effect as long as water remains ponded at an adequate depth between deliveries. For continuous irrigation, maintenance water flow rates are small. Intermittent irrigation utilizes larger streams for short periods. Intermittent irrigations with fluctuating water depth allow for storage of rainfall while still remaining continuously flooded.

Basins and non-level fields can be laid out in a variety of shapes and configurations for continuous flooding. Field sizes are limited primarily by topography and other practical conditions, not soil uniformity. Long dikes are made parallel whenever practical, as this facilitates mechanization and laser leveling. Management skill is required to establish dikes and control structures. Improperly placed dikes or control structures will reduce system performance. Dikes may be permanently constructed or annually broken down to facilitate harvest. In some areas they may be covered with plastic strips for erosion protection, seepage, and weed control. Wind-caused wave action may erode dikes and require extra maintenance.

Excellent land grading is very important. The fields may be graded, leveled as a basin, or have only a small elevation difference across the field to assure that an adequate (not too deep nor too shallow) ponded water depth can be maintained and the field drained for harvest. Figure 3-4 shows small scale continuous flood irrigation of a rice field which was not laser leveled. On larger low gradient tracts, laser leveling is often employed and in some cases where contouring is required, the dikes are installed with laser control. If such equipment is used, skilled maintenance is required. Earth moving associated with continuous flood systems is usually not more than 200 $m^3$/ha (100 cy/ac). When slope is left in the basins and/or the dikes are put on the contours, very little earth moving is required. Where larger basins are desired for easier farming, the grading requirement increases. Level surfaces may be essential for some germination procedures and for rotations using basin irrigation with other crops than rice.

Bench heights should be kept low so that water can be easily conveyed from one basin to another and to minimize washouts. This may limit locations where these methods can be economically applied. Many ancient fields have high benches on steep slopes and are not suited to current labor and mechanization conditions.

Figure 3-1.     Small Scale Continuous Flood Irrigation of a Rice Field.

Management of continuous flood irrigation involves pre-season land leveling, initial setup of dikes and water control structures, initial flooding of the field, and finally maintaining the proper water elevation in the flooded areas and assuring an adequate water supply with little waste.

Some means of surface drainage should be available to permit harvesting, and it is essential if rainfall occurs during the growing season on a heavy soil and a dry-footed (i.e., non-paddy) crop.

**Ponding (Fill and Drain)**

Ponding (fill and drain) is a variation of continuous flood irrigation adaptable to the same field conditions but the soil surface is intermittently exposed. Ponding can be used for any crop capable of being temporarily covered for the necessary irrigation duration.

Irrigation is accomplished by filling the diked area with water and holding the water until the desired infiltration depth has occurred. It is adapted to slow intake rate soils and poorly graded fields. It has been used on large 16 ha (40 ac) fields with as much as 0.3 m (1 ft) differences in elevations. Runoff control to drain away the ponded water is essential. Draining excess water through an outlet gate controls the desired irrigation amount. Duration of irrigation is critical for high application efficiency and, as with border strip irrigation, the advance and recession curves should be nearly parallel for high uniformities. With a long ponding period this aspect is not important, but it is critical with short periods used with light applications in bedded basins.

Ponding is adapted to heavy irrigations. It is applicable for leaching on sloping or poorly leveled small fields.

## Furrows

Furrows are sloping channels formed in the soil. Water moves down the furrow and infiltrates for a longer time at the upper end (time of application, $T_a$) than at the lower end (time of infiltration at the lower end, $T_l$) by the duration of the time of advance, $T_{adv}$, less the time of recession, $T_r$. The time of recession is relatively small on sloping fields, but can be significant on low gradients and soils with low intake rates. In order to assure adequate water and time at the lower end, runoff must occur, so return flow systems may be essential. Infiltration is often a slow process that occurs laterally and vertically through the wetted perimeter of the furrow.

Assuming that the soil is uniform and that good land grading has occurred, the three most important hardware items which facilitate simple and efficient furrow irrigation are (i) a tailwater return flow system which incorporates a reservoir, (ii) short furrows for an acceptable advance ratio, and (iii) a large variable water supply stream.

There are many furrow shapes; typically "Vee", trapezoidal, parabolic, and broad-flat shapes with wetted widths varying from 150 to 750 mm (0.5 to 2.5 ft) or more.

Furrows are well adapted to row crops and orchards or vineyards. They are less well adapted to field or other crops if cultural practices require tractor travel transverse to the furrows. Corrugations and shallow broad furrows can be used with flat field crops to alleviate some cultural problems. Furrows can even be adapted for wide-spaced vine crops such as melons or crops that should not lie on the wet soil. Vines may be "trained" out of the furrows, furrows adjacent to the plant may be relocated out or abandoned as plants extend, or alternate or widely spaced furrows away from the plant may be used as the root system expands (part area wet management).

Small, shallow seeded crop irrigations require careful land grading of the surface as overtopping furrows may cause crusting of the soil and high spots that do not become wet so that seeds do not germinate.

Furrows can be shaped and spaced to allow appreciable dry soil between them, to adapt to a great range of row spacings, or to create flat beds between them. Furrow spacing and shape is often determined by the spacing of equipment used for cultural activities and by ideal plant spacing.

The use of alternate side irrigation, applying water in alternated furrows on either side of a crop row at each irrigation, can be a desirable practice for row crops where higher frequency and smaller application depths are desirable. In this way, an individual crop row receives water at half frequency on one side or the other. Twice as large a set can be made with the same size stream. Labor needs can be minimized by adequate use of proper equipment and semi-automation of the supply. Alternate side irrigation are impractical if soils crack, as the water will move laterally into adjacent dry furrows.

The set time or application duration should be the time needed to infiltrate the desired depth of water at the low end of the furrow, plus the advance time, less the recession time if any. The spacing between the wetted edges of adjacent furrows should not exceed twice the distance across which water will move laterally by capillary movement within the duration of irrigation at the lower end. The entire surface between furrows may not appear wet but may be adequately wet below the surface. The distance between the wet edges can be adjusted by changing the shape as well as furrow spacing center-to-center. Such changes can greatly affect the intake rate and required irrigation duration. A typical furrow irrigation system with alternate row irrigation and siphon tube supply is shown in Figure 3-5.

The relative differences in time that water is at the upper end ($T_a$) and at the lower end ($T_l$) and the related infiltrated depths is described by the advance ratio. $AR = T_{adv}/T_l$. For non-cracking soils this ratio closely corresponds to the distribution of the applied water as to runoff, deep percolation, and stored at 100% Adequacy. Intake rate and desired duration of runoff at the tail end ($T_l$) are affected by changes in Managed Allowed Deficiency, furrow spacing, shape, compaction, reuse, slope, stream size, end blocking, and return flow systems. Uniformity and efficiency are modifiable by changes in intake rate and its many items, and especially stream size (AR). Runoff and deep percolation respond to changes in AR and the many items affecting it. As with most modern irrigation methods, to facilitate management it is extremely important that the supply system be flexible in rate as well as frequency and duration. Convenience and amount of required labor are related to the flow rate.

Furrow lengths usually range between 60 and 600 m (200 and 2,000 ft), but this range can be exceeded in very low intake rate soils unless root zone saturation (and associated root asphyxiation) occurs. Lengths are dominantly controlled by intake rates and stream size as well as field length. Furrow lengths may be shortened by use of supplemental cross ditches or gated pipe on continuous furrows. Use of large furrow streams relative to the furrow length reduces the excess time water is at the upper end and will result in more uniform infiltrated depths but more need for cutback streams or return flow systems to reduce runoff. The maximum stream size must not be erosive.

Figure 3-5. Alternative Row Furrow Irrigation with Siphon Tube Supply.

Intake rates in furrows are sometimes quite variable, even presuming that soils are uniform for the entire length. The variations are caused by cultural practices. New furrows have open soil conditions and high intake rates. Reuse of the same furrow after the soil has settled or slicked over will reduce the intake rate and assist in increasing the rate of advance. Driving a tractor wheel and/or dragging an object down a furrow will reduce the intake rate, and can make adjacent furrow infiltration more consistent, especially for new furrows. At the upper end of V-furrows, the stream size, wetted width, and hence, the intake rate, are greater relative to the lower end. In the broad furrow, there is very little difference between ends as the furrows are level across.

Intake rate also varies with soil and water temperature, cracking of soils upon drying, the magnitude of drying, cultural practices, and plow pans. In orchards, for example, the middle area between tree rows may have slower intake rates than the area adjacent to the tree row due to cultural operations. The rate of infiltration (expressed in dimensions of $LT^{-1}$) depends upon the percentage of wetted area -- a furrow on a wide spacing will have a different intake rate than a furrow in the same soil on a narrow spacing. The intake rate, if expressed in dimensions of $L^3/(LT)$ (e.g., Liter/Hour-meter), is affected by the furrow shape and not by the spacing.

Because of the numerous management controllable items, furrows can be adapted and modified for many conditions within the limits of soil uniformity and amenable topography. If the all conditions are just right it is possible to have high uniformity, and with a return flow system (Figure 3-6), a high efficiency is possible. However, because the uniformity and efficiency are highly dependent upon management, mismanagement can severely affect performance. The method can be used on short or long fields and with all but the most extreme intake rates, providing the rates are quite uniform. In general, the soil moisture depletion (MAD) needs to be greater than what is optimum for sprinkler and drip/micro systems.

Water is typically applied to individual furrows with siphon tubes (Figure 3-5), spiles, or through gated pipe. It may be done through cuts in the ditch bank. Guiding water to individual furrows may make furrow irrigation somewhat labor intensive, especially during initial irrigations or after cultivation. Using gated pipe with "sleeves" (socks) to guide streams and control erosion simplifies the work. Gated pipe furrow flow rates are adjustable and can be preset with practice.

Conventional furrows that have a fixed water supply may greatly restrict management's capability to optimize conditions. Fixed small steady flow rates as from a well or project supply make it difficult or impossible to obtain desirable advance ratios for acceptable uniformity and application efficiency if a large number of furrows are irrigated simultaneously. Therefore, these fixed small steady flow rates make it difficult to set large fields at one time to save labor, to make cut-back flows and to avoid runoff. Twenty-four hour fixed durations from a project supply, or small continuous flows from a well that needs to run much of

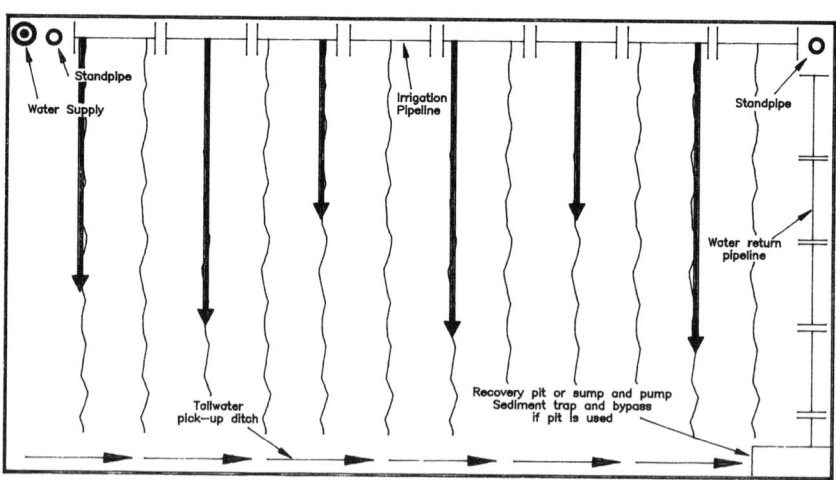

Figure 3-6.   Schematic of a Return Flow System Used in Conjunction with an Underground Distribution Pipeline.

the time to supply enough water, are not related to soil infiltration and intake rates and cannot be shut off when adequate water has infiltrated or be modified to daytime-only for convenience of labor and its effective use. Fixed frequency is often imposed by rotation schedules and results in lower yields affected by poor germination and failure to match crop water needs or to conserve rainfall.

As the result of these management constraints, conventional furrows are labor intensive, have only moderate efficiency values, and frequently are associated with mediocre yields. The labor-intensive nature of furrow irrigation, the need to control runoff, and the inadequately controlled extensive application of conventional furrow irrigation have led to numerous attempts at mechanization. The use of reservoirs, pipelines, siphons, gated pipe, socks, and irrigation runoff recovery systems (tailwater return systems) are the simple forms of mechanization.

Tailwater runoff is often restricted by efficiency concerns or by local regulations. Tailwater return systems are key elements for relatively simple furrow management. If tailwater is allowed to flow into a collection system, this reduces the irrigation management because irrigators do not need to spend as much time making sure all the furrows advance at the same rate. Furthermore, they do not need to spend time making many flow cutbacks. A problem with cutback flows is that typically irrigators start new furrows with the excess water, so that at the end of the day there are many furrows with many different application times. Ideally, a tailwater return system is designed in conjunction with the delivery system so that both the exiting (tailwater) and entering (supply) flows can be buffered. Also, tailwater return systems are easiest to manage if they have 12 hours or so of buffer storage, so that the water does not need to be used immediately.

There are several ways to reduce runoff through reducing the furrow onflow rate (i.e., cutback of inflow to the furrow) when the water has advanced to the end of the furrow and has run off for a short time. Using cutbacks can reduce the amount of storage needed for tailwater return systems. Two possible mechanical methods when the supply flow is not capable of providing cutback flows are cablegation and surge flow valves.

Cablegation is a special application of furrow irrigation. The concept of cablegation is shown in Figure 3-7. A plug restrained by a cable released at a constant rate from the upper end, is pushed through a sloping gated pipe by the water pressure behind the plug. As the water begins to flow to a furrow, the flow rate is high, with a fast advance. As the plug advances, opening new furrows, the head on the first outlet reduces and the flow rate decreases, yielding a gradually decreasing cutback stream until the flow stops. Thus, cablegation is an automation technique that can in theory significantly improve advance ratios while limiting runoff. In theory, this has a high potential application efficiency. Cablegation systems require considerable sidefall for proper operation, thereby

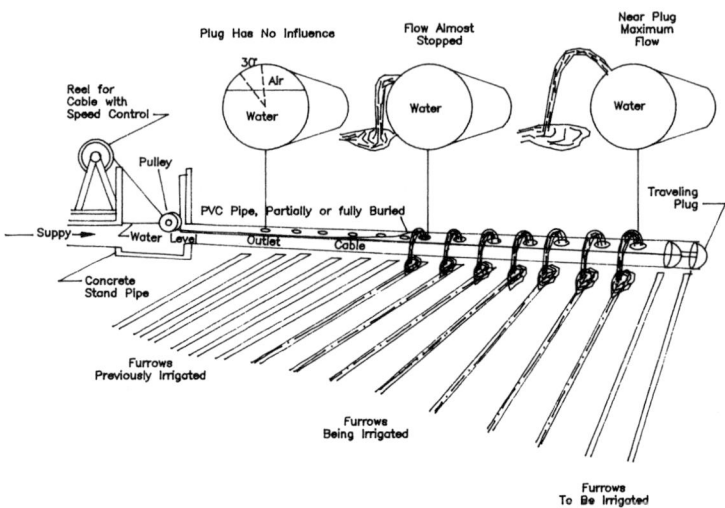

Figure 3-7.   Cablegation. From Kemper et al. (1985).

limiting their adaptability to sites with such conditions. The first and last groups of furrows in an irrigation block require special consideration. The method is rarely used because of its hardware and management complexity. It is interesting to mention because a key concept incorporated into it is the provision of cutback streams with limited labor.

Surge flow is a development for furrow irrigation that automates features of both the advance and the cutback stream processes. The concept of surge flow is demonstrated in Figure 3-8. The unique feature of surge flow is its repeated cycling of water from one set to another and then back -- typically with gated pipe. This process provides double-sized initial streams for rapid advance, lowers the water intake rate, especially on first surges by the process of wetting and recession, and ideally produces a more uniform distribution of water over the entire furrow length for some soils and does this automatically. These potential (although not always achieved) advantages are shown in Figure 3-8.

The surge flow system applies water in two phases. The advance phase consists of long flow periods to each section alternated with recession periods in which the furrows must become "dry". The ratio of flow time to recession time must be equal. During the advance phase, the alternate filling and drying of the furrow lowers the intake rate and smoothes the furrow surface facilitating subsequent rapid advance. Soon after the water reaches the end of the furrow after several surges, the cutback phase begins. In this phase, the surge cycle is very

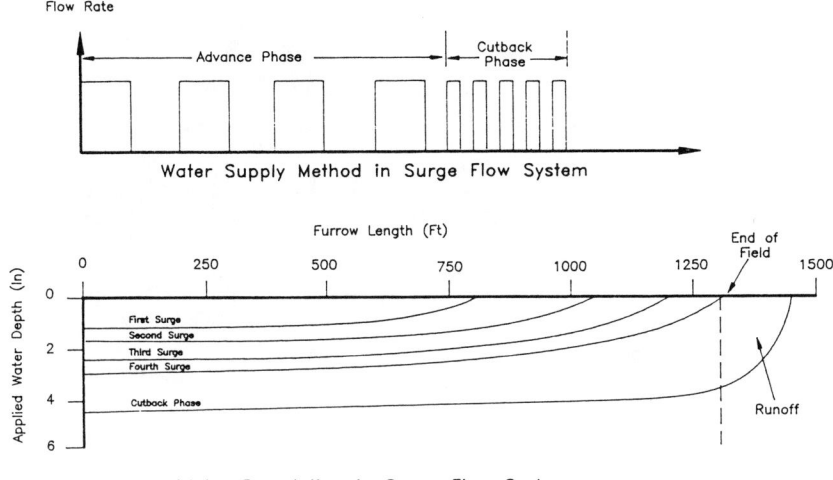

Figure 3-8.  Principles of Surge Flow. From Kubota (1986).

short (e.g., 5 minutes). There must be water in the furrow at all times to permit uniform infiltration along the furrow, but the time-averaged flow rate is half the initial flow rate, reducing runoff. Surge flow is capable of high uniformities up to 0.90 on some soils using half size supply streams. The sets must be small enough that the 2:1 cutback recession stream always has a little runoff, so the technique is not applicable to sandy soils with small intake rate changes with time.

Surge flow is made possible by alternating flows to two halves of a field. Gated pipe is used with a valve that diverts flow between the right and left sections of open pipe (Figures 3-9 and 3-10). Rugged solar-powered commercial surge valves are readily available. Sometimes farmers leave the valves in place and move the controllers between locations to reduce costs.

The acceptance of surge flow is highly variable in the U.S., depending upon geography. The most widespread applications have been in areas of summer rainfall such as Texas, where farmers have minimal tailwater runoff. In California there has been extensive research as well as extensive field trials, but in general farmers have not seen significant benefits to surge flow in most areas. In California, farmers typically achieve the same benefits by using shorter furrows and higher flow rates, with more simple management and hardware as compared to using surge flow systems.

Figure 3-9. Experimental Surge Flow System with Individual Valves at Each Furrow. (Such Designs Have Been Abandoned Because of Their Complexity and High Cost.)

Figure 3-10. Surge Flow System with Single Main Valve Diverting Water to Two Sections of Gated Pipe.

On-farm reservoirs, or their equivalent, can be used to accumulate small steady flows commonly available from irrigation districts or wells into large variable flows controllable at the point of application, and are the heart of some good surface irrigation systems (especially for furrows). These reservoirs at the head of the field can also receive tailwater runoff from a pump at the bottom of the field. As an example, over 250 farmer-built reservoirs have been installed in the Coachella Valley Water District of California to obtain flexibility for both the surface and pressure methods. In Coachella Valley as farmers have shifted to drip/micro irrigation, it has been discovered that these reservoirs are very important as de-silting basins. A reservoir for "overnight," 24, or 36 hour storage permits use of a return flow system and/or flexible use of a large irrigation stream accumulated from a smaller steady flow. Depending on topography, it can be used to regulate flows on 16 to 64 ha (40 to 160 ac). A medium sized 24-hour capacity reservoir of compacted earth construction would cost about $8000 for a small 15 ha (40 ac) farm and not much more for a larger farm utilizing the reservoir more days. A lined reservoir for similar areas might cost more than five times as much. Such a reservoir facilitates water, labor, and power conservation.

Numerous control techniques are available to modernize the water supply systems, both canals and pipelines, in order to provide more flexible deliveries. Those techniques are described in other literature, but it is essential to note that flexible water deliveries are important for all irrigation methods, including surface irrigation. For example, a very flexible water delivery system will allow farmers to cut back flows on a particular irrigation set at any time without needing to immediately increase the flow to other furrows (as would be required with an inflexible supply).

Land grading requirements for furrow irrigation are similar to those for border strip irrigation, except that some cross slope is not detrimental. Land planing with laser guided or conventional equipment is required for fields with very low gradients or slopes. For moderate to heavy grading, regrading is required after initial irrigation as the fill areas consolidate. Land grading becomes very important on low slopes (0.006 or less), to improve uniformity of irrigation. On slopes of 0.003 or less and especially for small seeded crops, precision land grading becomes essential since small variations in gradient are undesirable or even unacceptable, though on steeper slopes they are only mildly detrimental. Very small gradient fields may require annual land planing. Use of laser controlled equipment is very desirable. On steeper slopes, "contour" planting with some smoothing is practical for permanent crops such as orchards, but more difficult for annually replanted crops.

It is desirable that soils and especially intake rate be uniform for the full length of any furrow. They need not be uniform for an entire field as the difference across a field can be overcome by management. Appreciable soil differences cause non-uniform infiltration and should be alleviated by adjusting field shapes to conform to soil variations, or by over-cutting and replacing top soil to provide reasonable uniformity when doing land grading. The economic impact of such extensive grading or irregular field boundaries must be considered. Other systems may be more adaptable to non-uniform soils.

Although soils of all intake rates can be irrigated by furrows, the extreme rates (high or low) require excessive labor or capital and are seldom economical. The extreme soils may also have low productivity with furrow irrigation. Excessively high intake rate soils will typically have high deep-percolation losses. Low intake rate soils will usually have high runoff and long duration sets.

Labor requirements for water distribution in furrows range from low to high. They are related to the degree of mechanization and to stream size and its controllability. Simple sloping earth ditches with turnouts are very cheap but labor intensive. Permanent ditch checks, concrete lined ditches, and surface or buried pipeline result in reduction of labor. Use of siphons and particularly gated pipe also reduces labor. Regardless of the degree of automation or mechanization, training and experience is necessary to achieve uniform irrigations. The uniformity of irrigation is highly dependent upon experienced operation of the system.

## **Corrugations**

Corrugations are a variation of furrows. The same principles of operation and automation apply. They are usually smaller, run straight (or nearly so) downhill, and are more closely spaced. They may be constructed before or after the crop is established. They may serve to guide shallow depths of water downslope in close growing crops in fields with some cross slope. They can be used on steeper slopes in conjunction with short lengths as they utilize small non-erosive streams. The method is most adaptable to irregular topography and steeper slopes where land grading is not practical or economical and only land smoothing is needed. It is difficult to implement modern scientific irrigation practices with corrugations.

Corrugation irrigation has the same flexibility as furrow irrigation in terms of stream size, with the same labor restrictions. Because the corrugations are smaller than furrows and the fields are usually smaller, smaller flow rates are usually used.

The land grading requirements for irrigation with corrugations is less than other surface methods. Steeper downrow slopes are possible and some irregularity in downrow slope is not detrimental. While some cross row slope is tolerable it cannot be large due to the small size of the corrugations.

## Contour Ditches (Wild Flood)

With this method, ditches are constructed nearly on the contour or on ridges and water is spread by flooding the field downslope from the ditch. Little or no land grading is employed but corrugations may be added. Appreciable land may be left dry. Ditches are usually earthen with portable dams and cutouts through the ditch bank used to distribute the water to the field. Capital cost is low, labor requirement is moderate to high or negligible, depending on the level of control achieved. Uniformity is low. The method is usually employed in areas where water is inexpensive and may be unreliable (e.g., mountain valleys with high snow melt runoff in the spring and early summer and low base flows with no storage). When available the water is spread on low value, erosion resistant crops such as grass raised for pasture or hay. When water is not available the crop, goes dormant. Unless a cover crop already exists, crop establishment can be difficult due to erosion until cover develops, requiring careful control of the water and a great deal of labor. This is a very unscientific method of irrigation.

## CAPABILITIES AND LIMITATIONS

The primary advantages of surface irrigation methods are:

1. They typically use very simple irrigation equipment, and if equipment fails, water can often be applied if enough hand-labor shovel work is employed.

2. On many soils and topographies, these methods have the lowest initial capital investment.

3. Silty and dirty water can be used, where filtration of that water might be very expensive for sprinkler or drip/micro.

4. High efficiencies and uniformities are possible given the correct combination of conditions such as medium-heavy soil types, excellent land grading, uniform soils, excellent management, a large variable flow rate supply, and tailwater (runoff) return systems.

5. In many cases, surface irrigation systems require no pumps. This advantage, however, can also be considered a negative because tailwater return systems require pumps, and tailwater return systems are essential for the effective management of most furrow and border strip systems.

6. If the systems have flexible and large flow rate supplies and tailwater return systems (for sloping methods), they can have low labor requirements.

The primary disadvantages of surface irrigation methods are:

1. They require the most "art" of all the irrigation methods, both to obtain a high DU and a high AE. In general, people have not learned the art.

2. The DU of surface irrigation methods is extremely sensitive to soil differences within a field.

3. It is difficult to apply small depths of water evenly with a high AE.

4. Irrigation scheduling, on a scientific basis, is difficult and requires excellent historical records on each field.

5. Excellent land grading is required for some of the surface irrigation methods. It is difficult to obtain excellent land grading on small fields.

6. The efficient furrow systems of the western U.S. that utilize tailwater return systems and have excellent land grading are difficult to economically duplicate in the very small fields found in other areas of the world.

## **Crops**

Virtually all crops have been grown using surface irrigation methods. Surface irrigation methods are best adapted to Soil Moisture Depletions (SMDs) of 5 cm or greater if a good DU and high AE are desired; some crops grow best with lower SMDs and ideally would be irrigated more efficiently with other methods.

## **Soils**

Surface irrigation methods are most successfully used on medium-low intake rate soils. Many clay loam and silt loam soils have medium-low intake rates at the end of the irrigation season, but they may have very high intake rates early in the season. For this reason, farmers may irrigate with sprinklers early in the season and then switch to furrows later when the intake rates are lower.

## **Topography**

In general, surface irrigation methods are suitable for slopes of less than 1%. Steeper slopes are irrigable but generally some degree of water control is lost and erosion (with accompanying siltation of rivers and streams) can be a serious problem. While contour furrows are intended for steep slopes, many of these situations in the U.S.A. have been converted to drip/micro or sprinkler irrigation methods.

**Water Supply**

Previous sections have pointed out the importance of having a flexible and reliable water supply. If the irrigation project or well does not provide a flexible supply, on-farm reservoirs can be constructed to provide flexibility and to also facilitate simple and efficient use of tailwater return flows.

**Salinity/Water Quality**

Water quality and salinity are generally not a problem for paddy irrigation. Some through-flow and seepage typically keeps the water salinity low. Water salinity can become dangerously high if there is an upward flux of water and there is no through-flow of water.

For surface irrigation methods with beds, salinity damage can be minimized by using special bed shapes such as creating a ridge in the center of the bed to which the salt will "wick up". Planting is done on the edges of the bed.

Basin systems provide excellent opportunities for salinity and leaching control. Ponding of water on the surface can leach salts out of the root zone. All the surface methods can leach the majority of the root zone with sufficient time. However, furrow beds and border strip ridges cannot be leached effectively. In many areas with salty water, sprinklers are used for the initial germination and pre-irrigation events, and later furrow irrigation is used.

An advantage of all the surface irrigation methods is that the plant leaves are not contacted by salty water, as is the case with most sprinkler irrigation methods.

**Climate**

Surface irrigation methods are generally incapable of applying small depths of water uniformly across a field, except during the latter part of the growing season when the intake rates of the soil may have significantly decreased. Therefore, it is difficult to manage surface irrigation methods with precise small water applications that would allow a farmer to leave part of the root zone dry enough to store rain. An option used by some farmers is to use the alternate side technique in the event that when it rains at least half of the field will have the ability to store some rainfall.

Surface irrigation prior to frost is a common practice to achieve 0.5 - 1.5°C frost protection in orchards in California. The protection comes from a darker soil (increased retention of incoming solar radiation during the daytime) and from the water acting as a buffer and providing a heat source during evenings.

In some areas, farmers with paddy irrigation note that the water provides a temperature buffer during very hot days. They have less rice damage with paddy irrigation than with intermittent irrigation methods.

Evaporation from wet soil surfaces before crop maturity is greater in the full coverage methods (e.g., basin, border strip, continuous flood) than those with only partial wetting (furrow, corrugation bedded basins).

Surface irrigation methods have an advantage over many sprinkler methods in that they are insensitive to wind conditions.

**Efficiency**

Surface irrigation methods have a broad range of potential application efficiencies that are affected by topography, flexibility of water supply, soil conditions, boundary constraints and management. Under ideal conditions, most methods, when well designed and well operated, can have high $PAE_{lq}$'s. However, there are nearly always constraints that cause conditions to be less ideal. Therefore, the practical $PAE_{lq}$'s will be somewhat less. Table 3-2 presents the most likely $PAE_{lq}$'s for the various methods under ideal and practical conditions.

Table 3-2  Potential Application Efficiency for Surface Irrigation Methods Under Ideal and Practical Conditions.

| Method | $PAE_{lq}$ Ideal | $PAE_{lq}$ Practical |
|---|---|---|
| Basin | 80% | 75% |
| Border Strip | | |
|    Sloping w/runoff | 85% | 75%[1] |
|    Low Gradient/Blocked | 90% | 80% |
|    Contour Border | 80% | 70%[1] |
|    Contour Ditch | 50% | 40% |
| Continuous Flood | 85 | 80 |
| Ponding | 80 | 75 |
| Furrow | | |
|    Traditional Sloping | 75% | 70%[1] |
|    Mechanized | 85% | 75%[1] |
|    Contour | 75% | 65% |
| Corrugation | 75% | 65% |

[1] with irrigation runoff recovery systems, increase by 10-15%

The actual application efficiency ($AE_{lq}$) will nearly always be lower than the $PAE_{lq}$ due to management error. Some of the surface methods are more sensitive to management than others. For comparison purposes, average or above management levels are assumed. Since there can be significant variation in efficiency due to local conditions, discussions of the effect of these conditions on efficiency for the various methods follow.

The potential distribution uniformity ($DU_{lq}$) at 100% Adequacy for all the surface irrigation methods will depend upon a combination of factors that can be grouped into two categories: (i) opportunity time differences, and (ii) intake rate differences.

Different soil-water contact times (opportunity times) are caused by:

- differences due to advance and recession of the water throughout a pond, basin, furrow, or border strip,

- differences in application time and advance/recession between ponds, basins, furrows, and border strips, (which are typical when comparing daytime against nighttime sets), and

- differences in land grading.

Intake rate differences are caused by

- different soil types and tilth throughout a field,

- different amounts of compaction throughout a field, such as between wheel rows (rows which are consistently compacted by tractor tires) and non-wheel rows,

- different amounts of salinity throughout the field, especially sodium, and

- different wetted perimeters of furrows within a field.

Most irrigation system evaluations have historically only considered differences in intake opportunity times along single furrows, border strips, etc. (i.e., AR), and therefore have overestimated the actual field $DU_{lq}$.

## **Irrigation Scheduling**

Scientific irrigation scheduling is more complicated and difficult with surface irrigation methods than with drip/micro and sprinkler. The soil (rather than sprinklers or emitters) controls the rate of water infiltration during an irrigation set, making it important that the SMD match the amount of water that one expects to infiltrate at the low quarter point during an irrigation event. Because both the target SMD and the intake rate will vary from one irrigation event to another, efficient irrigation scheduling requires careful observation and documentation of irrigation practices and infiltrated depths on each soil and for each irrigation event. After a history is built up, one can estimate how fast water will advance and recede, and how much water will infiltrate at the LQ point for a typical irrigation at a particular time of the year.

Irrigation scheduling is complicated by the fact that each time the furrow spacing is changed, or the soil is tilled, intake rates will change. Also, early in the season it may be impossible to apply a small depth of irrigation water uniformly, due to high intake rates. Although soil intake rates can be manipulated by changing furrow shapes, by using surge flow, using soil compaction or tractor wheel compaction or torpedoes, the results are not always predictable.

In short, surface irrigation scheduling is much more "artful" than for sprinkler or drip/micro irrigation, but with careful attention to historical records, the art factor can be reduced.

## INSTITUTIONAL CONSIDERATIONS

### Labor

Labor is discussed in the descriptions of each surface irrigation method. Labor can be very un-sophisticated if water is simply spread around a field. As one progresses to modern surface irrigation techniques, labor must be trained to understand the various factors that impact DU and application efficiencies.

In the U.S., the irrigators are often blamed for poor performance of surface irrigation methods, but upon closer inspection the blame often belongs on their supervisors. Efficient border strip and furrow irrigation, for example, typically require a flexible irrigation supply and a good tailwater return system. If these are lacking, the irrigators have severe limitations placed upon them and it is almost impossible to achieve both a high DU and a high AE.

Labor can be minimized by using large flow rates, good land grading, a flexible water supply, tailwater return systems and good water delivery hardware.

### Service Availability

Access to good land grading equipment and persons with good experience in land grading is the most critical service component. Beyond that, access to surge flow valves, low pressure pipelines, and basic valves is important. In general, the hydraulic equipment is simple to install and repair.

# ECONOMIC FACTORS

## Capital Costs

Construction costs for surface irrigation systems are highly variable depending on grading requirement and water distribution system costs. Nearly all methods require some type of land grading, although it may be minor for some methods and site conditions. Some means of delivering water to the head or high side of the field is needed. This may be in earth ditches, concrete ditches or pipelines. In some cases an on-farm reservoir may be required to allow local control of the water delivery schedule. If automated equipment is required, there will be an additional cost. Return flow systems are mandatory in many areas to prevent tailwater runoff from farms. They permit the use of large streams and semi-automation with conservation of water and labor. Many different configurations of tailwater return flow systems exist; local topographies, existing flow rates, and proximity of the adjacent fields must be considered.

Table 3-3 lists typical unit costs for the various components of construction typical in large scale agriculture in the U.S. Local site conditions, equipment availability, and farm size will influence these typical costs.

The cost per unit area for a completed system will depend on the field layout and size, soil intake rate, amount of grading required, and the type of water delivery system. Table 3-4 presents the range of typical costs for systems in the United States utilizing the unit costs from Table 3-3. These costs are meant as a general guide only. The range in actual field conditions can be much larger. Costs are so site specific that individual cost estimates should be completed for each system analyzed.

Table 3-3.  Unit Costs for Surface Irrigation Components.

| | Unit Costs | |
|---|---|---|
| Item | Metric | (English) |
| land grading | $1/m$^3$ | ($0.75 yd$^3$) |
| laser finish | $145/ha | ($60/ac) |
| earth ditch | $4/m | ($1.25/ft) |
| concrete ditch - 1 m | $69/m | ($21/ft) |
| concrete ditch - .5 m | $39/m | ($12/ft) |
| pipeline - .4m | $39/m | ($12/ft) |
| pipeline - .6m | $98/m | ($30/ft) |
| large turnout | $1,200 ea | ($1,200 ea) |
| small turnout | $120 ea | ($120 ea) |
| gated pipe - .15 m | $8/m | ($3/ft) |

Table 3-4.  Estimated Construction Cost Ranges for Various Surface Irrigation Methods.

| Category | | Basin | Border | Contour Ditch |
|---|---|---|---|---|
| | | **Metric Units** | | |
| Land grading volumes | m³/ha | 400 - 1600 | 200 - 1200 | 0 - 100 |
| Land grading costs | $/ha | $370 - $1,480 | $190 - $1,110 | $0 - $90 |
| Laser finish | $/ha | $150 - $150 | $150 - $150 | N/A |
| Border/berm construction | $/ha | $100 - $100 | $70 - $70 | N/A |
| Supply system | | | | |
| Earth ditch | $/ha | $440 - $590 | $170 - $640 | $100 - $490 |
| Concrete ditch | $/ha | $2,070 - $3,420 | $590 - $2,400 | N/A |
| Pipeline | $/ha | $2,450 - $4,450 | $1,330 - $2,400 | N/A |
| Add for gated pipe | $/ha | N/A | N/A | N/A |
| | | **English Units** | | |
| Land grading volumes | yd³/ac | 200 - 800 | 100 - 600 | 0 - 50 |
| Land grading costs | $/ac | $150 - $600 | $75 - $450 | $0 - $38 |
| Laser finish | $/ac | $60 - $60 | $60 - $60 | N/A |
| Border/berm construction | $/ac | $40 - $40 | $30 - $30 | N/A |
| Supply system | | | | |
| Earth ditch | $/ac | $180 - $240 | $70 - $260 | $40 - $200 |
| Concrete ditch | $/ac | $840 - $1,385 | $240 - $970 | N/A |
| Pipeline | $/ac | $990 - $1,800 | $540 - 970 | N/A |
| Add for gated pipe | $/ac | N/A | N/A | N/A |

Table 3-4. (Continued)

| Category | | Continuous Flood | Furrow | Corrugations |
|---|---|---|---|---|
| | | **Metric Units** | | |
| Land grading volumes | m³/ha | 100 - 400 | 200 - 1200 | 100 - 400 |
| Land grading costs | $/ha | $100 - $370 | $190 - $1,110 | $100 - $370 |
| Laser finish | $/ha | N/A | $150 - $150 | N/A |
| Border/berm construction | $/ha | $70 - $70 | N/A | N/A |
| Supply system | | | | |
|   Earth ditch | $/ha | $170 - $640 | $170 - $640 | $170 - $640 |
|   Concrete ditch | $/ha | $370 - $1,040 | $540 - $2,150 | $540 - $2,150 |
|   Pipeline | $/ha | $540 - $1,110 | $540 - $2,150 | $540 - $2,150 |
|   Add for gated pipe | $/ha | N/A | $120 - $490 | $120 - $490 |
| | | **English Units** | | |
| Land grading volumes | yd³/ac | 50 - 200 | 100 - 600 | 50 - 200 |
| Land grading costs | $/ac | $40 - $150 | $75 - $450 | $40 - $150 |
| Laser finish | $/ac | N/A | $60 - $60 | N/A |
| Border/berm construction | $/ac | $30 - $30 | N/A | N/A |
| Supply system | | | | |
|   Earth ditch | $/ac | $70 - $260 | $70 - $260 | $70 - $260 |
|   Concrete ditch | $/ac | $150 - $420 | $220 - $870 | $220 - $870 |
|   Pipeline | $/ac | $220 - $450 | $220 - $870 | $220 - $870 |
|   Add for gated pipe | $/ac | N/A | $50 - $200 | $50 - $200 |

## Energy Cost

Energy costs are highly dependent upon the source of energy supply, the source of water, the method of water conveyance between fields, and the topography. It is not unusual for 60 ha fields using aluminum gated pipe to require booster pumps delivering 210 kPa (30 psi). Likewise, it is common to have ditch systems with no pumps involved. Effective tailwater return systems typically require pumps.

## Labor Cost

Labor cost for operation of surface irrigation systems is highly variable, depending on system design, field size, supply stream size and degree of automation and whether day time only operation is practical. Table 3-5 can be used as an estimator for labor requirements for the various surface irrigation methods under average conditions, with the understanding that local conditions

Table 3-5.  Labor Requirements for Surface Irrigation.

| Method | Labor Required per Irrigation | |
|---|---|---|
|  | Man-hrs/ha | Man-hrs/ac |
| **Basin** | | |
| large scale | 0.25 | 0.10 |
| small scale | 1.25 | 0.50 |
| **Border Strip** | | |
| standard | 0.65 | 0.25 |
| guided or contour | 1.25 | 0.50 |
| contour ditch | 6.20 | 2.50 |
| **Continuous flood or ponding** | | |
| large scale | 0.25 | 0.10 |
| small scale | 1.25 | 0.50 |
| **Furrow** | | |
| traditional sloping | 4.95 | 2.00 |
| traditional w/siphon tubes | 1.25 | 0.50 |
| traditional w/gated pipe | 0.25 | 0.10 |
| mechanized | 0.10 | 0.05 |
| contour | 2.45 | 1.00 |
| **Corrugation** | | |
| earth ditch | 6.20 | 2.50 |
| siphon tube | 1.85 | 0.75 |
| gated pipe | 0.50 | 0.20 |
| mechanized | 0.10 | 0.05 |

may cause the actual labor requirement to vary. A large flexible water supply requires very little labor as one man can handle irrigation of 200-400 ha if a large tailwater return flow system is also provided. Typically, the labor costs will be highest in the first year or two while the system stabilizes and will then reduce somewhat with time.

**Operation and Maintenance Cost**

Operation and maintenance cost (other than irrigation labor) for surface irrigation systems consists mainly of periodic re-leveling, reinstallation of secondary ditches, borders, berms, gated pipe, or furrows and maintenance of the water delivery system. Typically, releveling of precision leveled fields for basin, border strip and furrow irrigation occurs every two to three years for annual crops, at a cost of about $150 per ha ($60 per ac).

Maintenance of field structures (secondary ditches, borders, berms and furrow) may occur multiple times per year for crops that are cultivated between irrigations. For semi-permanent and permanent crops, they would occur primarily when the crop was reestablished or maintenance was required due to deteriorating conditions.

Maintenance of field ditches, pipe, outlet structures and control structures is required periodically to preserve the integrity and capacity of the system. The costs will depend upon the type and quality of the ditches and structures being maintained. Generally, earth ditches and control structures in them will require the greatest maintenance and pipe systems will require the least.

As a general guide to maintenance cost, the values presented in Table 3-6 may be used. Local experience should be used to refine these values when available.

Table 3-6. Typical Operation and Maintenance Costs for Surface Irrigation Methods.

| Item | Basin | Border | Contour Ditch |
|---|---|---|---|
| | | $/ha/yr | |
| periodic releveling | 49 - 74 | 49 - 74 | |
| field structure maintenance (borders, furrows, etc.) | 10 - 12 | 10 - 12 | |
| earth ditch maintenance | 25 - 37 | 25 - 37 | 25 - 37 |
| concrete ditch maintenance | 12 - 17 | 12 - 17 | |
| pipeline maintenance | 5 - 7 | 5 - 7 | |
| | | $/ac/yr | |
| periodic releveling | 20 - 30 | 20 - 30 | |
| field structure maintenance (borders, furrows, etc.) | 4 - 5 | 4 - 5 | |
| earth ditch maintenance | 10 - 15 | 10 - 15 | 10 - 15 |
| concrete ditch maintenance | 5 - 7 | 5 - 7 | |
| pipeline maintenance | 2 - 3 | 2 - 3 | |

| Item | Continuous Flow | Furrow | Corrugation |
|---|---|---|---|
| | | $/ha/yr | |
| periodic releveling | | 49 - 74 | 49 - 74 |
| field structure maintenance (borders, furrows, etc.) | 10 - 12 | 10 - 12 | 10 - 12 |
| earth ditch maintenance | 25 - 37 | 25 - 37 | 25 - 37 |
| concrete ditch maintenance | 12 - 17 | 12 - 17 | 12 - 17 |
| pipeline maintenance | 5 - 7 | 5 - 7 | 5 - 7 |
| | | $/ac/yr | |
| periodic releveling | | 20 - 30 | 20 - 30 |
| field structure maintenance (borders, furrows, etc.) | 4 - 5 | 4 - 5 | 4 - 5 |
| earth ditch maintenance | 10 - 15 | 10 - 15 | 10 - 15 |
| concrete ditch maintenance | 5 - 7 | 5 - 7 | 5 - 7 |
| pipeline maintenance | 2 - 3 | 2 - 3 | 2 - 3 |

# CHAPTER 4
# DRIP/MICRO IRRIGATION

## DESCRIPTION

Drip/micro irrigation (also previously referred to as "trickle" irrigation) refers to a variety of irrigation methods in which water is delivered directly to small areas adjacent to individual plants through emitters or applicators placed along a water delivery line. In an orchard or vineyard there will typically be one or more emission devices per tree. A schematic of a typical drip/micro system for orchards or vineyards is shown in Figure 4-1. Figure 4-2 shows a microsprayer system in an almond orchard.

For row crops (e.g., broccoli, lettuce, peppers, cotton) and field crops (alfalfa, grains) the emission devices are spaced closely enough so that the capillary action of the soil provides water to each plant root zone. It is unusual to use drip on broadcast field crops because of the difficulties in wetting all of the plants and the low prices of such crops. A schematic of one subsurface row crop drip irrigation system design is shown in Figure 4-3. Everything is buried except the block valves and the flushouts and air vents. Block valves are shown in Figure 4-4.

In the irrigation industry the preferred terminology to describe these systems varies by individual and geography. However, "drip" irrigation generally refers to systems which use low flow rate emitters from which water drips onto the soil. "Micro" irrigation often refers to systems with emission devices that throw water horizontally and vertically with a spray or sprinkler pattern. Some people in industry distinguish between "microsprayers" that have no moving parts, and "microsprinklers" that have rotating parts. Drip/micro irrigation systems are almost always "solid set", meaning that equipment such as hoses and emission devices remain in one place during the growing season. Systems may be permanently installed (typical for trees and vines and for some row/field crops) or may be portable and moved to a different field after an irrigation season is completed. Other systems are hybrid, with a buried mainline distribution system and removable or disposable laterals and/or manifolds/submains.

Drip/micro irrigation systems require very clean water to avoid plugging of the emission devices. Filtration components represent a major portion of the cost and maintenance of drip/micro irrigation. In addition, chemigation is generally required to avoid plugging due to bacterial growth and/or chemical precipitation in the laterals and emission devices. Flow rates for individual emission devices are typically very small, although some microspray systems have such large nozzle diameters that they might also be classified as small flow rate permanent, solid set sprinklers. Typical ranges of emitter flow rates are provided in Table 4-1.

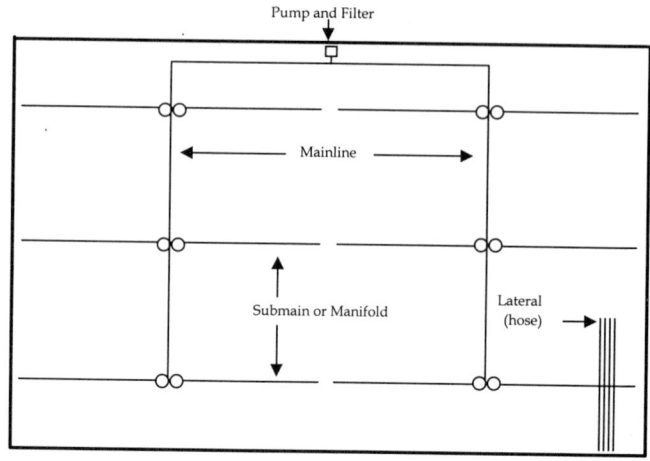

Figure 4-1.  Schematic of Typical Drip/Micro Irrigation System on Trees or Vines.

Figure 4-2.  Microspray System on a Newly Planted Almond Orchard. Bakersfield, Calif.

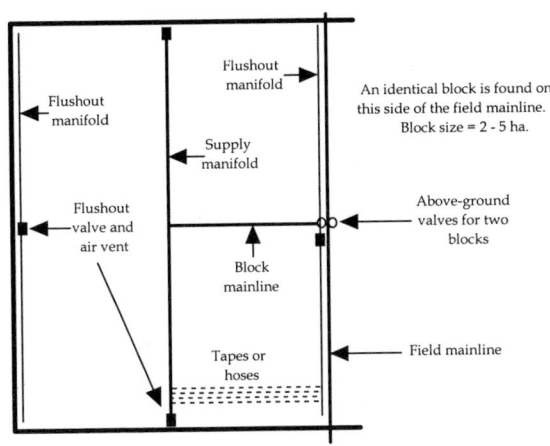

Figure 4-3.   Schematic of a Subsurface Row Crop Drip Design Typical of the Central Coast of California.

Figure 4-4.   Block Valves for a Permanent Subsurface Row Crop Drip Design.

Table 4-1.    Flow Rates of Typical Emission Devices in Drip/Micro.

| Emitter Type | Typical Range of Flow Rates (LPH) |
|---|---|
| Row crop drip tape | 0.5 - 1.2 |
| Row crop drip hose with discrete emitters | 1.2 - 3.6 |
| Vineyard/orchard drip emitters (above ground) | 2.5 - 11.0 |
| Vineyard/orchard drip emitters (below ground) | 1.3 - 3.0 |
| Vineyard/orchard microsprayers/microsprinklers | 22.0 - 75.0 |

Because drip/micro irrigation systems are "solid set", many of them are easily automated. They are ideal systems for irrigation managers who are interested in fine tuning the applications of water and fertilizer (fertigation) through the irrigation system. Irrigation water is generally applied through emission devices daily or several times per week. Some managers pulse the systems on hourly intervals, although that practice is not standard. Some of these irrigation systems are designed to irrigate a whole field at once. However, the trend toward higher emitter flow rates (such as microsprayers) or more closely spaced emitters (such as on row crops) usually requires that the full pump flow rate be rotated between two to eight blocks within a single field.

**TYPES OF DRIP/MICRO IRRIGATION**

There are many variations of drip/micro irrigation systems. Some of the differences are due to agronomic or horticultural requirements. For example, frost protection is very important for citrus and avocados in some regions, and micro sprinkler/sprayer designs offer better climate control than do emitters. Drip emitters may be preferred in almond orchards because they enable one to irrigate alternate tree rows without wetting the soil around adjacent rows, as would happen with microsprinkler/sprayer designs. This alternate row irrigation is important with almonds because alternate rows may be planted with different varieties that require stress at different dates prior to harvest. An orchard crop with an extensive, shallow root system such as avocado will typically perform better under microsprinkler/sprayer than under drip. Conversely, closely spaced (hedgerow spacing) trees are better suited to drip emitters, because there are so many emitters in such a design, the wetted soil volume is high, and microsprayer/sprinkler designs suffer from problem of tree and trunk interference of the sprayer patterns. Citrus growers in some regions prune the trees so that the leaves never touch the ground; microsprayers in these situations can wet a large area. If the citrus is pruned so that the leaves touch the ground, microsprayers in effect become high flow rate drip emitters because the water hits the leaves and cannot spread out. Drip emitters typically wet less soil area per emitter on sandy soils than on loam or clay soils, given the same water quality. Therefore, it is more expensive to use drip on sandy soils than on heavier texture soils because more emitters (and sometimes an extra hose per tree row) are needed on sandy soils to obtain sufficient soil wetted area (often desired to be in the 60% range). Microsprayer/sprinkler systems would cost the same on either soil type, because the wetted area is so large that the capillary spread of water beyond the spray pattern is not very important.

**Orchard/Vineyard Drip (Above Ground)**

These systems typically have one hose per plant row on closely spaced rows (row spacing less than 4 meters), and may have two or more hoses per row on wider spaced rows. Emitters are often spaced in arid regions so that at least 60% of the potential root zone volume is wet, which provides an adequate moisture reservoir for periods of high evapotranspiration, and as insurance against several days of breakdowns. Less wetted area is common in areas with supplemental rainfall.

The emitters used in orchards and vineyards are generally manufactured separately from the hose, and they may be installed on the hose either at the factory or in the field, depending upon the emitter configuration and design. Most hose is manufactured from polyethylene, with common diameters of 16 - 30 mm. Hose lengths (from the hose inlet) vary from about 100 to 200 meters. In the case of orchards, a single hose is generally installed down the tree row, on the soil surface, right next to the tree trunks with only a small percentage of extra length (1.5 - 2.5%) to accommodate hose expansion and contraction due to temperature changes. Recent designs rarely use spaghetti tubes to move water from the emitters to distant locations, although this was common in earlier designs. The use of spaghetti tubes has been discontinued because it was found that after time, the spaghetti tubes were typically wind-blown or kicked together. If a single line of emitters will not provide sufficient soil wetted area, it is common to install two hoses, one on each side of the tree row but out of the way of tractor traffic.

On vineyards a single hose per row is almost universal because the rows are tightly spaced. Usually one or two emitters per vine are used. Depending upon the region and harvesting/tillage equipment, the hose may be placed on the soil surface next to the vine trunks, or be suspended in the air at a height of approximately 0.3 meters (see Figure 4-5). Suspension requires the existence of a trellis system with wires onto which both the vine branches and the hose are attached. Suspension provides the ability to till under the vines without damaging the hose and emitters. There has recently been some interest in using 2 hoses per vine row, with each hose supplying one emitter per vine. By alternating the irrigation between the hoses (for example, one hose for 2 weeks and the other hose for another 2 weeks), the growers are able to alternately stress different sections of the root system. Some growers feel that this alternate stressing can reduce transpiration and increase grape quality without sacrificing grape tonnage.

Figure 4-5.   Suspended Hose and Emitter on a Vineyard.

Orchard and vineyard drip systems were well established on large acreages in many areas of the world by the early 1980's. The equipment has continued to improve, with excellent choices now available of well designed emitters and hoses. Most emitters are now of one of two designs, tortuous path or pressure compensating. Tortuous path designs are popular because they provide relatively large passageways and reduced plugging problems compared to vortex or laminar flow emitter designs. Tortuous path designs also provides a reasonable degree of pressure compensation (flow rate changes are approximately proportional to the square root of pressure changes). Because they have no moving parts, emitters with tortuous designs tend to be relatively inexpensive, well made, and durable.

The second most popular emitter design incorporates some type of pressure compensation provided by a moving part that progressively restricts the passageway size as the pressure increases. There are many brands and models of such pressure compensating emitters available, but only some are robust and have the desired characteristics when new and for many years later. Some have excellent self-flushing characteristics. To minimize problems with these types of emitters, purchasers should specify a very low manufacturing coefficient of variation (manufacturing CV typically less than 0.05), and a guarantee that the pressure/discharge relationship follows a pre-defined curve. A well-written and backed warranty should also be obtained regarding those same factors (CV, average discharge at a specific pressure, and how the flow changes with pressure) after 3 - 5 years of operation.

## **Orchard/Vineyard Drip (Subsurface)**

Buried drip systems on orchards and vineyards are a relatively new concept, with limited acreage but they are the subject of a considerable number of popular and research articles at the time of this publication. The theoretical reasons for farmer interest in subsurface drip irrigation are clear - they include less soil evaporation, fewer weeds, less humidity in the orchard (and therefore less disease problems), and the ability to drive and till throughout a field at any time, regardless of the irrigation schedule.

In practice these advantages are often not achieved because the water may rise directly to the soil surface rather than slowly spreading through the soil around the emitter by capillary action. Causes include too high flow rates per emitter and low hydraulic conductivities of the soil. The low hydraulic conductivity is often affected by the irrigation water quality. Solutions include treatment of water quality problems (with polymers such as polyacrylamides, or gypsum injection), using very low flow rate emitters, and pulsing the irrigation system automatically in durations of less than 30 minutes to 1 hour. The requirement of using low flow rate emitters is contrary to industry history that has shown a trend towards larger flow rates to minimize plugging problems. The pulsing can also cause non-uniformity of water application during start up and shutdown times, unless the hoses are short and the blocks are small and on flat terrain. Some new emitter designs have a built in "check valve" feature that stops emitter drainage once the hose pressure drops to around 2 psi (14 kPa), which might be useful on fairly flat ground.

Catastrophic failure can occur due to roots pinching the hose, or due to root intrusion into the emitters. At the time of this publication, the only proven technique available to avoid root intrusion is to use emitters that are impregnated with Trifluralin, or to periodically inject Trifluralin. Any injection of Trifluralin in the U.S. must follow labeling restrictions.

An additional uncertainty with buried drip systems is the proper depth and location of hoses. Depths vary from 0.45 - 0.75 m, with the shallower depths more common. The location must be such that wheel compaction of the soil over the hoses does not occur. On established vines, some growers place the hose midway between the rows. In orchards, growers are experimenting with many configurations, including a buried hose midway between tree rows and an above-ground hose on the tree row itself. Caution must be taken when installing buried systems on established crops, as extensive root damage may occur (see Figure 4-6).

Figure 4-6.   Roots Which Were Uprooted During Installation of Buried Drip on Established Pistachio Trees.

## **Orchard/Vineyard Microspray (and Microsprinkler)**

Microspray and microsprinkler systems (called "micro" in this section) became very popular in the early 1980s, and many drip systems were converted to micro at that time. In many cases, it is unclear which system is better; drip or micro. Micro systems typically have larger hose diameters than drip because the flow rates of the emissions devices are much higher than for drip. They also tend to have smaller hose lengths than drip for the same reason. Because of the high application rates, a micro field is often divided into 6 or more blocks with only one being irrigated at a time, whereas many drip systems are only divided into two blocks. The net result is that micro systems are more expensive than drip systems. The exception would be on widely spaced plants such as walnuts, in which case several drip hoses would be required per tree row compared to only one hose for micro.

Microsprayers and microsprinklers are typically connected to the lateral hose with a spaghetti hose of 0.3 - 1.0 m in length. This allows the lateral hose to move due to temperature changes or equipment contact, yet the emission devices remain standing undisturbed. There are many designs of stakes to support the microsprayers and microsprinklers above the ground (typically 0.1 - 0.3 m high) (see Figure 4-7).

Figure 4-7.    Microsprayer During Irrigation.

An introduction to advantages and disadvantages of micro versus drip is given in the beginning of this section. Micro has an apparent advantage of requiring less stringent filtration than drip because of the large and short paths of micro nozzles, resulting in less potential for plugging. However, this might be deceiving because the high nozzle velocities, accompanied with silt particles (which are not typically removed by filtration) can cause abrasion of the nozzles and spray plates. Abrasion causes an increase in flow rate and a distortion of the spray patterns. Some growers with good filtration (150 mesh and better) now only assume a 6-8 year life for the nozzles themselves. The actual life will depend upon the polymers used in the nozzles and spray plates, as well as the type of contaminant and degree of filtration.

Micro can also provide some frost protection, plus gives a larger soil wetted volume than a single hose drip system. Frost protection is achieved in some areas by actually placing the microsprayer in the citrus canopies during periods of frost. The microsprayer is re-positioned on the ground for the irrigation season. The use of micro for frost protection should be done with extreme caution, as a severe frost can quickly convert the microsprayer discharge from a protective device to an evaporative cooling device.

The disadvantages of micro, as compared to drip, include the higher cost in some designs, the higher evaporation losses (if the water is extended past the canopy), and inability to easily restrict the wetted area during certain times of the year. Some emission devices are pressure compensating. Pressure compensation may be done in the emission device itself or with the use of special flow control barbed fittings at the inlet of the spaghetti hose. Caution should be used when selecting such devices, since the quality of some of these devices is poor.

One product on the market contains an accumulator at the base of the microsprayer. A low flow rate supply fills up a chamber, and once full, the chamber rapidly empties out through the microsprayer. This pulsing feature enables one to convert an existing, low flow rate emitter design into a microsprayer system with an intermittent high flow rate per microsprayer.

## **Row Crop Drip (Above Ground)**

Above ground row crop drip irrigation has been in existence since the earliest years of drip. Presently there are four major categories of above ground drip.

1. In Florida's coral sand soils, the drip hose is typically part of a "plastic culture" in which drip tape (thin walled hose with integral emitters built into the walls or seams of the tape) is placed under plastic as it is installed for vegetables. The surface location is important because there is very little capillary action in the soil to provide upward movement of water from buried emitters. The emitter flow rates of these above ground systems installed under plastic sheets are relatively high.

2. Many growers of pole tomatoes, sugar peas, and similar crops use disposable drip tape products for one or two seasons. These growers often have small fields that are very difficult to irrigate by any other means due to the small flow rates available and uneven field sizes. The above ground drip systems provide an easy way to provide frequent irrigation on these high value crops; they also eliminate the problems of wetting the foliage and fruit as would occur with sprinklers. These systems, due to their short life, often have under-designed filtration systems and minimal maintenance. The tape is typically thin-walled (4 - 8 mil).

3. Certain crops such as celery are not well suited to buried drip because of the harvesting conditions or rooting systems. However, their yields and crop quality can be enhanced with the continuous high moisture content which drip provides. Celery harvest is characterized by heavy equipment criss-crossing moist soils, which would compact or destroy any buried tape. Sweet potatoes have aggressive root systems that tend to give more root intrusion problems with buried drip than do other truck crops. On other crops, some growers prefer a totally portable system for various reasons.

Therefore, growers often use retrievable tape or other drip hose with internal emitters that can be rolled up from the end of the field. Commercially available equipment has been developed to retrieve the tape or hose for reuse. The extent of reuse is often limited by the large number of couplers that must be used in the hose as it is reused many times on fields with slightly different row lengths. Work has been done on in-field hose splicing techniques to reduce the need for cumbersome couplers. At the time of this publication, no effective permanent in-field splicing techniques are available.

4. The greatest growth area of above ground row crop drip is for a wide variety of short crops (broccoli, lettuce, peppers, onions, etc.) which are not adversely affected by a wet soil surface (such as tomatoes would be). Farmers have discovered the advantages of drip, but do not want to invest in the very intensive management and special equipment that are necessary for the more permanent buried drip. Surface drip wetting patterns are less sensitive to soil differences than with buried drip. Removable tape also allows for different crop row spacings during crop rotations. It is common to place the tape about 1 - 5 cm below the ground surface - just enough to protect it from the wind - and then retrieve it immediately before or after harvest. Some of this tape is used for 3-10 seasons before disposal. Submain/manifolds may be permanent or portable.

## Row Crop Drip (Subsurface)

There are two main categories of subsurface row crop drip: "one crop" and "permanent". They are described below

1. "One crop" buried drip systems have almost dominated the irrigation of strawberries and sugar cane in the U.S. and Hawaii since the 1970s. These systems typically use a permanent buried mainline and submain system, along with permanent filtration systems. The tape is buried 10 - 25 cm before, during, or immediately after the crop is planted or transplanted. The buried tape has less problems with wind and tillage damage than above ground tape would have.

   In the case of strawberries, drip tape is ideal because the strawberry beds are typcially raised high and covered with plastic so that the berries will not contact wet soil and suffer from mold. The drip system allows frequent irrigation of the shallow, sensitive root system and keeps the fruit dry.

   Sugar cane, due to its height, is unsuitable for the use of hand move or side roll sprinkler systems. Center pivots and linear/lateral move systems can clear the crop with special high tower machines, but the uneven field sizes and very rough terrain often preclude their use. Furrows cannot be used effectively because of the extensive and vigorous growth of the cane that blocks off the furrows. Rough terrain also rules out furrows in many areas. For these reasons, drip tape has been found to be an excellent irrigation choice for sugar cane.

For strawberry irrigation systems, the tape and plastic mulches are generally picked up and disposed of after harvest and before the plants are disked into the soil. For cane systems, the tape cannot be removed prior to harvest, so it remains in the soil unless it is burned along with the foliage prior to harvest. In both strawberry and cane systems, new tape is typically used for the next crop.

Asparagus may fit into this category, although it is a long term crop. There is some concern about root intrusion with asparagus, so many growers use emitters with a slow release herbicide incorporated into the plastic.

2. Permanently buried drip on row crops. These systems have become increasingly popular in the southwestern U.S., with major expansion of acreage since 1991. As of 1998, there are approximately 150,000 acres of this type of irrigation in the U.S. (rough estimate by the first author), with a high degree of interest among vegetable growers and researchers. The primary crops grown with these systems are high value such as tomatoes, peppers, lettuce, cauliflower, and broccoli. Dozens of other crops have been placed into crop rotations with these systems, including wheat. The systems generally have permanent water filtration systems (see Figure 4-8), mainline, and submain components. The drip tape or hose is buried 20 - 40 cm below the ground surface and is designed to remain in place for 6 - 10 years. Special tillage equipment is required to remove the old crops and incorporate the crop residue into the soil without damaging or moving the tape. Successful systems generally have owners who are interested in total system management, including sophisticated fertigation techniques. The systems require a very high level of management skill and attention. During the first year or two of operation, a farmer must often spend a very high percentage of total management time dealing with the just one field of this type.

In order for the buried drip system to provide adequate irrigation during germination or transplanting, special design features must be incorporated, raising the cost of the drip system itself. Typically the design must have emitters of high flow rate (about 3 LPH per meter of hose), relatively thick tape walls, and adjustable pressure regulators at the entrances to the blocks. The pressures are increased during germination or transplanting, and the high flow rates may actually

Figure 4-8.   Media Filtration for a 60 ha Row Crop Drip System.

saturate the soil surface and cause runoff. During the rest of the season, the system is operated with lower emitter pressures so that the soil surface remains relatively dry. If these systems are not capable of providing germination and transplant irrigations, temporary portable sprinkler systems must be brought into the field at those times. Use of a sprinkler system represents additional costs and time delays in establishing a new crop.

In addition, there has been some interest in using permanent buried drip systems on traditional field crops such as corn, especially in states such as Kansas. These systems have typically used fairly deep burial (40 cm or so) and very wide emitter spacings (150 cm). It is expected that continued research will indicate what spacings and depths are most applicable for this supplemental irrigation, and if it is economically viable.

## CAPABILITIES AND LIMITATIONS

Drip/micro irrigation systems have the following typical advantages over some other irrigation methods:

1. They can be used effectively on extremely steep ground.

2. They require minimal land grading. Land grading is necessary, however, to prevent surface drainage problems which might occur with rain, and to accommodate any special tillage equipment used.

3. It is more difficult to have gross over-irrigation during months of peak ET. This is generally because many drip/micro systems are not designed with a large system pump capacity.

4. The Distribution Uniformity (DU) of new systems can be very high (0.93 or greater) in reasonable terrain and with an excellent design, because the new system DU depends only upon hydraulics and equipment design, rather than upon management, soil differences, and/or overlap patterns of sprinklers. As the industry provides even better pressure compensating emitters, the new system DU values will probably consistently fall in these high (>0.93) ranges.

5. Systems can be installed on virtually any size or shape of parcel.

6. Generally, there are no runoff problems to contend with.

7. The systems are capable of high frequency irrigation (without a degradation in DU [such as occurs surface irrigation methods] or excessive non-beneficial evaporation losses [such as occurs with sprinkler methods]). High frequency irrigation allows the maintenance of an optimum soil moisture content in the root zone, which is especially important for salty water, or for shallow rooted crops. It should be noted that very high frequencies are not necessarily optimum for some crops such as lettuce.

8. Fertilizer can be directly applied uniformly to the root zone at any stage of growth on any day and with any dosage, without wetting plant foliage (see Figure 4-9).

9. The upper portion of the root zone can be maintained moist, which enhances the uptake of nutrients, such as phosphorus and ammonium, that are typically concentrated near the soil surface.

The above advantages are "typical", meaning that they are not universal. As with any other irrigation method, good operation depends upon good design, good equipment, and good maintenance. Similarly, drip/micro irrigation systems can have the following disadvantages:

1. The DU can degrade quickly with time due to insufficient water filtration, lateral flushing, and/or chemical injection. In other cases, the DU may degrade quickly due to some unforeseen and unusual circumstance such as fresh water clams growing inside the hoses, or an

Figure 4-9.    Fertigation Tanks for Row Crop Drip.

unusual insect that prefers to nest inside a certain type of emitter. Rodent damage can be devastating in some areas.

2. These systems are susceptible to damage by vandalism, and vandalism repairs may be complicated, time consuming, and costly.

3. Evaporation losses can be high with some micro designs that frequently wet large areas of bare soil.

4. Although the potential for excellent results (water savings, fertilizer efficiency, optimization of yield) often exists, they can only be achieved with excellent design and excellent management. Often it takes several years for irrigators and farmers to develop even average management skills, and catastrophic failure can result before those skills are gained. Drip/micro methods are sometimes perceived by growers and planners as being the "silver bullet" or magical cure for problems associated with other irrigation methods that are actually caused by the poor management with those other methods. Unless the management style changes when drip/micro systems are installed, the problems may change from being moderate to being severe. Fortunately, many growers are willing to completely re-learn irrigation operation techniques when introduced to a new irrigation method.

5. Water must be available to the system on a very frequent and dependable basis. Drip/micro cannot be used in irrigation projects that deliver water on a rotation schedule (which includes most of the world's acreage) unless the fields are supplied with groundwater.

6. Energy costs for installation and operation of the irrigation system itself are usually higher than for surface irrigation methods (assuming similar efficiencies) on flat ground. However, the total energy use efficiency may actually be higher under drip/micro if fertilizer usage is reduced and if yields are improved; similarly, energy requirements may be less under drip/micro if less land grading is required.

7. There are dozens of different types of essential parts (fittings, valves, etc.). The systems must be supported by an excellent resupply infrastructure.

8. The initial cost of some forms of drip/micro, in particular some permanent row/truck crop systems, is among the highest of any irrigation method.

9. In very arid areas, a sprinkler system may be needed once every few years (and in some cases more often, such as with some buried row crop drip) to leach salts which have built up near the soil surface.

## **Crops**

The preceding paragraphs show that the crop type often dictates the type of drip/micro system that will be used. In addition to factors already mentioned, disease control is often a critical concern. Spray patterns of microsprayers are often oriented away from tree trunks to minimize various types of trunk and root rot, especially on heavy soils where maintaining adequate aeration is a challenge. Maintenance of the proper orientation of the sprayers over several years can be problematic. Drip has been especially successful with some crops such as tomatoes and peppers because phytophthora problems on heavy soils are much less with drip than with sprinklers or furrow irrigation. Peppers are grown under drip in some areas of heavy soils where farmers found it impossible to achieve an economical yield in the past due to the disease problems. Some crops such as pasture and hay have not yet been drip irrigated successfully on a large commercial scale. They may eventually be irrigated successfully if the wheel traffic of the harvesters can be confined to repetitive patterns that do not cross over the buried hose/tape. Drip/micro has been successfully and economically used on high value crops such as fruits and vegetables. Economic success has been more elusive with lower value field crops.

Drip/micro irrigation of any crop almost always requires a period of learning for a farmer who may have grown that crop for years under sprinkler or surface irrigation. Melons, for example, seem to respond much quicker to a drip irrigation than they do to a furrow irrigation. Care must be taken to not overirrigate some such fruit crops, because they can quickly turn vegetative.

## Soils

Drip/micro irrigation has been used on virtually all soils - from coral sands to heavy cracking clays. The soil type will impact the number of emitters used per plant, as well as the decision to use micro vs. drip. On trees in very sandy soils, micro would be the choice over drip because of the limited lateral movement of water under drip.

## Topography

Drip/micro is found on virtually any topography. It is the most adaptive of irrigation methods in this regard.

## Water Supply

There are two primary considerations regarding water supply for drip in addition to the universal need for enough quantity during a season - flexibility of water delivery and water quality. Drip/micro irrigation systems require the availability of almost continuous flow rates to a field during the periods of peak evapotranspiration (ET). As with sprinklers, the supply must be flexible enough to adjust the hours of operation per day or per week at all times of the season to match the ET rate. Furthermore, if automation is contemplated, the water supply must be available on true "demand". As mentioned earlier in this chapter, the majority of the irrigation projects in the world supply water on rotation, which makes implementation of drip/micro almost impossible unless ground water is used as a supply instead of the surface supply.

Reservoirs are often used to buffer the irrigation supply, whether it be an irrigation district supply or from wells. The flow rate from well pumps may change during a year as groundwater levels fluctuate, yet a drip/micro system needs a constant flow rate into a block. By discharging the wells into a reservoir and then boosting the water out of that buffer reservoir, the problem can be solved.

## Salinity/Water Quality

Drip/micro systems can typically use saltier water than other irrigation methods because drip/micro can maintain the soil moisture at a high optimum water content, thereby reducing osmotic stress. However, the emitters are very sensitive to solids concentrations in the water. Extensive and expensive filtration is needed for dirty source water (Burt and Styles, 1994). In some cases, the water is so dirty that reservoirs are needed to settle out sand and silt, or to oxidize iron in well water. Those reservoirs serve as pre-filtration, prior to the regular filters. Drip/micro systems also are sensitive to fairly low (greater than 0.2 ppm) concentrations iron and manganese, as these can cause plugging problems unless chemically treated.

## **Climate**

One of the advantages of drip irrigation is that water is applied directly to the soil. Therefore, wind has no effect on water distribution. Microsprinklers and microsprayers typically have very low profiles and are within orchards, so wind effects are minimal. Evaporation losses may be less than, equal to, or greater than those found in other methods. They depend upon the frequency of irrigation, the percentage of the soil wet, the type of soil, and the location of the wet soil compared to shade. Typically, evaporation losses on most deciduous trees are considered to be higher under drip/micro than for surface irrigation (10 - 15% compared to 3 - 5%).

## **Efficiency**

Irrigation and application efficiencies are impacted by three major components: (i) Distribution Uniformity, (ii) actual duration of irrigations as compared to needed duration, and (iii) unrecovered losses such as evaporation and uncollected runoff. An analysis of drip/micro system performances shows that there are some inherent weakness and advantages. Ultimately the efficiency must be treated the same as for any other irrigation method. Good design, installation, and management pay off with optimal efficiencies. Poor design, installation, and management pay off with undesirable efficiencies. New drip/micro systems can generally be designed with a new system $DU_{lq}$ of .88 - .94. These new $DU_{lq}s$ can be guaranteed by a designer, and are among the highest consistently available $DU_{lq}s$ available for irrigation systems. However, actual $DU_{lq}s$ for older systems in the field are consistently lower, with numerous well documented studies in California (Little, no date; Cachuma RCD, 1994; Mission RCD, 1993; ITRC 1997-99) showing that the average $DU_{lq}$ of drip/micro systems (.70 - .85) is very similar to the system DUs of other irrigation methods. Primary causes of the lower values are flow variation due to poor emitter design, plugging, and pressure differences within fields. The conclusion is that very high field DUs can be obtained and maintained only with proper design, installation, and maintenance, and that most drip/micro systems do not qualify as such in the field.

On the efficiency side, drip/micro systems generally have application efficiencies that are higher than those for other methods. This is partly due to the ability to schedule irrigations at almost any time and for any duration, and partly due to the built-in flow rate limitation of most drip/micro pump designs.

## Irrigation Scheduling

Drip/micro irrigation systems are among the most simple to schedule accurately. They are capable of applying both small and large amounts of water with the same uniformity, as compared to surface irrigation methods that tend to have very poor DU's with low application amounts. Furthermore, the scheduling is very simple because the hours of irrigation are easy to adjust, and the infiltrated amounts for most cases depend only upon the hours of application, not upon the soil intake characteristics. The result is that the simplicity of scheduling typically leads to savings in applied water amounts as compared to prior irrigation methods, even though the DU may be the same.

Some farmers unknowingly under irrigate because they expect large water savings, and the annual irrigation efficiencies on drip/micro irrigation fields are above 90% because of the small deep percolation amounts and gross under irrigation on parts of the fields. A good water management plan would recommend that these farmers improve the system DU and reduce the under irrigation, and the desired annual irrigation efficiency would approach 80 - 85% or so rather than some value greater than 90%.

As noted in other sections of this chapter, the large percentage of continuous wetted soil surface with many drip systems and microspray systems (contrary to many published descriptions of drip/micro) typically results in higher evaporation losses from the soil surface than would occur with other irrigation methods. Depending upon the percentage of soil wet, duration of wetting, and the degree of plant shading, non-beneficial evaporation can vary from about 4% to 20% of the applied water on mature orchards and vineyards. With row crop drip, high evaporation losses can occur during the transplant/seeding stages (depending upon the degree of soil wetting and if sprinklers are used), but can be almost zero after the crop has sufficient cover. Permanent subsurface row crop drip systems are usually managed to wet only a minimum surface area once the seeding/transplant stage has passed.

Runoff is not normally considered to be a problem with drip/micro systems, but it can be a serious problem in some areas with water penetration problems. Although the percentage of runoff is typically low in these areas (less than 5%), the localized runoff patterns within an orchard or vineyard cause serious problems with equipment movement.

## INSTITUTIONAL CONSTRAINTS

### Labor

With drip/micro irrigation, a manager must be very high on the learning curve when the system is first installed, because there is very little time for learning and there is a small margin for error. This contrasts with other irrigation methods such as hand move sprinkler or furrow irrigation, for which the problems are usually not so difficult to solve, and can be worked out almost with brute force before a crop failure occurs. Good training and high levels of motivation are essential for all layers of drip/micro irrigation system management, from the farmer to the repair person.

Permanent subsurface row crop drip systems require the highest level of sophisticated and manual labor during the first installations on a farm. This is because these systems do more than irrigate differently; they require a completely different way of farming. For example, the tillage equipment, fertilization techniques, and irrigation scheduling are all very different from those of conventional farming. Farm laborers must be taught to pay attention to very fine details such as maintaining precision tracking while conducting tillage operations, and not crossing over row ends with tractor tires. Picking crews, who are often not employed directly the farm, must also be aware of special needs of the shallow buried laterals.

**Service Availability**

A major challenge in some areas is the unavailability of qualified designers, irrigation dealers, and installers. Because drip/micro irrigation systems have so many parts that must work together flawlessly, they are very sensitive to good design and installation. Furthermore, it is essential that spare parts of all types be readily available, and that qualified personnel be available to install the replacement parts. A sophisticated drip/micro irrigation system is destined to fail unless spare parts and local maintenance expertise are very readily available. Drip/micro irrigation systems are less forgiving than many other irrigation methods (except pivots and linear/lateral moves) in regards to irrigation scheduling. Typically the systems are not designed with a lot of reserve capacity, so once a soil moisture deficit builds up it is very difficult to recover and refill the root zone. Irrigation managers often anticipate large water savings with drip/micro, and they are unaccustomed to managing such systems; there is a strong tendency to under irrigate at first with resulting production decreases, and root intrusion on buried drip systems.


## ECONOMIC FACTORS

**General**

The discussions below focus on the expense side of an economic analysis of drip/micro irrigation systems. An economic analysis must also focus on the income, or reduced expense, aspects of a project. Drip/micro systems have been very cost effective on millions of acres because of labor savings, fertilizer savings, water savings, and improved yields or quality of yields. Of course, such numbers are not transferable from one location to another or from one grower to another. Drip/micro has also seen its share of catastrophic failures due to some unforeseen problem or due to poor design, installation, or management.

There are two different situations to consider. One is the conversion of an existing system to a drip/micro irrigation system. The other is a choice of what method to use on a new piece of ground. A conversion analysis has the benefit of data regarding the performance of the existing system. That performance can be compared to "potential" performance in the area, and the potential difference can be compared to the additional costs involved in a system conversion. Sometimes conversions are not economical because the existing investment must be abandoned.

For a new piece of ground, drip/micro irrigation systems often have an economic advantage because basic land preparation costs can be much lower than for surface irrigation. The fields can also be laid out in an optimum configuration to minimize system costs.

**Capital Costs**

Capital costs in the U.S. are generally much lower than in many developing countries because design services and equipment are readily available in the U.S., with minimal shipping charges. The costs given below are for U.S. systems. Design costs are real and must be paid for, whether the cost is imbedded in an irrigation dealer bid or directly charged by a consulting engineer. Design costs will vary from about $75 - $300/ha, depending upon the amount of information available and the size and complexity of the project and number of fields. Those costs can be greater if water supply information is lacking and if groundwater supplies must be investigated.

The purchase of a drip/micro irrigation system has similarities to the purchase of an automobile. While some people believe that air conditioning and power windows are essential in an automobile, others will be satisfied with a used car having neither. Similarly, there is no one correct specification of performance details that is applicable to all drip/micro systems. However a drip/micro irrigation system should always be purchased with a clear knowledge of what it contains and what the performance (DU, warranty, flow rate, pump TDH, etc.) will be. The point is that there is a wide range of capital costs for any type of irrigation system, including drip/micro. There is usually some tradeoff between initial costs and pumping costs, as well as future maintenance and performance. Costs are also highly dependent upon the spacing of the plants or plant rows. For example, a microspray system for a widely spaced walnut orchard will be much less expensive than one for a vineyard, which has many more rows (i.e., hoses) and plants (i.e., microsprayers). Approximate initial cost ranges (material, installation, tax) are found in Table 4-2.

Table 4-2. Approximate Initial Costs for Drip/Micro Irrigation Systems in the U.S. (Excluding Design).

| System Description | Cost, $/ha ($/ac) |
|---|---|
| Drip system on vineyards | 2000 - 3200  (800 - 1300) |
| Drip on orchards - surface | 1500 - 2700  (600 - 1100) |
| Drip on orchards - buried | 2000 - 3500  (800 - 1400) |
| Micro on orchards | 1800 - 3000  (700 - 1200) |
| Above ground row crop drip | 1000 - 2500  (400 - 1000) |
| [annual hose replacement costs] | [350 - 750    (150-300)] |
| Subsurface row crop drip, permanent | 2300 - 6300  (900 - 2500) |

## Energy Costs

The pumping cost is directly proportional to the cost of power, the irrigation efficiency over a season, the Total Dynamic Head (TDH) of the irrigation system, and the crop irrigation water requirement. One advantage of many drip/micro systems is that because they are "solid set", they can often be designed and automated so that they can be operated only during periods of the day/week with reduced electrical rates. Taking advantage of these lower "time-of-use" electrical rates does not reduce the energy consumption, but it can reduce pumping bills by a third or more.

Annual irrigation efficiencies of drip/micro irrigation systems tend to be high because of inherent limitations to gross over-irrigation related to low-medium system flow rate capacities. The TDH of drip/micro systems for flat terrain tends to be about 280 - 310 kPa (40 - 45 psi) for vineyard and orchard systems, and 210 - 280 kPa (30 - 40 psi) for row crop systems. The TDH requirements are dependent upon the type of filters required and selected. Energy audits for the California Energy Commission on drip/micro systems (Styles and Burt, 1996) found that a well designed drip/micro system often has a higher energy efficiency than do other systems because of two major reasons: (i) reduced fertilizer applications, and (ii) higher yields.

## Labor Costs

It is almost impossible to define labor costs, because they are so highly dependent upon the design, the type of crop, and the quality of installation. Furthermore, they are extremely dependent upon the attitude, sophistication, and management style of both the owner and operators. A very high performance (high DU, minimal plugging) requires a minimum of labor, but it absolutely requires that the system be installed correctly, with the proper filtration, flushout valves, and chemigation system. Furthermore, it requires a level of understanding that is more sophisticated than needed for most other irrigation methods, and a close attention to meticulous, periodical maintenance. If this is done, the labor will be minimal but systematic. There are farms in the U.S. of over a thousand hectares of trees with only one operator, with occasional requirements for a repair crew. On the other extreme, a system with serious rodent problems and also with a poor design, insufficient filtration, and inattention to chemigation may have one person working full time on a 100 ha field in the U.S., while maintaining a minimal performance (DU of about .60). "Smart" labor which involves periodic and necessary maintenance of automated filters and chemical injection is essential; without it, the regular maintenance requirements can be huge.

## Management Costs

For permanent subsurface row crop drip systems, it is not unusual during the first season for a manager of a farm with 20 fields to spend 30 - 40% of his time on one drip irrigated field trial. Some farmers believe that unless this is done, successful expansion of the acreage cannot occur. Delegation of authority and learning does not seem to work well with these systems during the early years.

## Operation and Maintenance Costs

As with labor, O&M costs are highly variable. Beyond the normal requirements of a good design, equipment, and installation, some other factors can arise. Surprises frequently arise in new drip/micro installations, puzzling even veteran designers and farmers. Examples of such "surprises" include wasps that lay eggs in microsprayers of a certain configuration but not in other configurations, birds that remove emitters of a certain color, sand from wells with unusual densities that cannot be easily removed by sand separators or media filters, wireworms that bore through drip tape, and microscopic slimy snails that live in wells and cause filters to plug. Such surprises can be expensive, and should be expected for initial installations. Generally, they can be solved over time.

# CHAPTER 5
# SPRINKLER IRRIGATION

## DESCRIPTION

In sprinkler irrigation, water is applied to the soil using a pressurized piping system with nozzles, jets, or perforated pipe that sprays the water into the air. These sprinkler devices or perforations are spaced to give a relatively uniform application of water over the field being irrigated using a series of sets or a continuous move system. A variety of sprinkler devices are available. These devices may be peculiar to a specific type of system but often are adaptable to a number of systems.

**Rotating Head Sprinklers**

Rotating head sprinklers apply water in a circular shaped pattern around the sprinkler nozzle location. The diameter of the circle to which water is applied depends on the sprinkler nozzle design and trajectory angle, size of the nozzle orifice, and water pressure at the orifice. The depth of water applied is normally greater near the sprinkler and decreases as the distance from the sprinkler increases.

Uniformity of application is accomplished by spacing the sprinklers so that the wetted area of adjacent sprinklers adequately overlap, usually about 60 to 80 percent. The sprinkler manufacturer will normally provide performance data for each type of sprinkler, nozzle size, and nozzle pressure which gives the discharge rate and wetted diameter. This data is essential in selecting sprinklers and designing the system for proper sprinkler overlap. Application patterns may be checked in the field after system installation. Poor application uniformity occurs around the periphery of the field where there is no adjacent set of sprinklers to overlap. This may be overcome by the use of part circle and tipping sprinklers along the edge of the field. Special non-overlapping heads are used in orchards where tree interference prevents uniform overlapping patterns.

Water pressure has a significant effect on the operation of sprinklers. Pressures higher than the recommended will break the water up into finer drops, causing more water to fall near the sprinkler and slightly reducing the area of coverage. Inadequate pressure results in a donut shaped pattern with excess water near the perimeter. Neither pattern overlaps effectively to provide uniform coverage. The slightly flattened top conical pattern resulting from moderate pressure for the nozzle diameter required will give overlap patterns that achieve uniform coverage.

Wind will distort sprinkler application patterns. The higher the wind velocity, the greater the distortion. Fine spray is more susceptible than large droplets to wind distortion. Wind conditions must therefore be considered in designing and operating the sprinkler system. Generally, this is done by reducing spacing and pressure and by using low angle sprinklers.

Rotating head impact sprinklers are often classified according to the pressure, related to nozzle size, and are required to apply the water in a recommended overlapping uniform pattern. The Irrigation Association has grouped impact sprinklers as follows:

1. Low-pressure sprinklers (35 to 200 kPa) (5 to 30 psi),

2. Intermediate-pressure sprinklers (200 to 400 kPa) (30 to 60 psi),

3. High-pressure sprinklers (above 400 kPa) (above 60 psi),

4. Large-volume sprinklers (above 500 kPa) (above 80 psi).

Burt and Keller (1977) further separated the low-pressure group to show the performance of very-low pressure sprinklers (35 to 140 kPa) (5 to 20 psi). A similar classification, along with the adaptability of each classification, is given by the USDA Natural Resource Conservation Service. For all sprinklers, the maximum point application rate is greater than the design average rate.

1. Low-pressure sprinklers (35 - 100 kPa) (5 - 15 psi)

    a. Wetted diameter - 6 - 15 m (20 - 50 ft).

    b. Average design application rate - 10 mm/hr (0.4 in/hr).

    c. Adaptation - Soils with intake rate in excess of 12 mm/hr (0.5 in/hr). Often used in orchards.

2. Moderate-pressure sprinklers (100 - 200 kPa) (15 - 30 psi)

    a. Wetted diameter - 20 - 25 meters (60 - 80 feet).

    b. Average design application rate - 5 mm/hr (0.2 in/hr).

    c. Adaptation - All field crops and vegetables.

3. Intermediate-pressure sprinklers (200 - 410 kPa) (30 - 60 psi)

    a. Wetted diameter - 25 - 35 meters (80 - 120 feet).

    b. Average design application rate - 6 mm/hr (0.25 in/hr).

  c. Adaptation - Field crops, overhead sprinkling of orchards.

 4. High-pressure sprinklers (410 - 690 kPa) (60 - 100 psi)

  a. Wetted diameter - 35 - 70 meters (110 - 230 feet).

  b. Average design application rate - 12 mm/hr (0.5 in/hr).

  c. Adaptation - Field crops, overhead sprinkling of orchards.

 5. Hydraulic or giant sprinklers (550 - 825 kPa) (80 - 120 psi)

  a. Wetted diameter - 60 - 120 meters (200 - 400 feet).

  b. Average design application rate - 16 mm/hr (0.65 in/hr).

  c. Adaptation - Close growing crops with good ground cover.

  For rotating head impact sprinklers, special application nozzles are available that maintain distance of throw for intermediate pressure sprinklers while operating at moderate pressures. These are often referred to as "controlled droplet size" (CDS) or "diffuser" nozzles. Other nozzles, termed "flow control" nozzles, are designed with a flexible orifice that reduces the open area as the pressure increases, maintaining relatively constant discharge over a wide range of pressures. Combinations of diffuser and flow control capabilities are also available, allowing relatively uniform discharges at low pressure, even in fields with relatively large elevation differences. These nozzles can be used without the pressure loss associated with pressure regulators.

**Low Pressure Spray Nozzles**

  Low pressure fixed spray nozzles spray water in a relatively horizontal plane, usually in a half or full-circle pattern, and are often used on center pivot and linear move systems or in orchards. The full-circle device causes the water jet from a nozzle to impinge on a plate, producing a spray in all directions. The plate may be flat, concave, convex, or complex depending on the distribution pattern desired. The plate may be fixed or may move as the water impinges upon it, either in slow rotation or by "wobbling". These moving plate systems provide somewhat greater throw distance and larger droplet sizes. Smooth plates produce a fine mist while grooved plates produce large droplets. For the half-circle pattern, the jet is deflected by a curved surface instead of a plate. The low pressure nozzles operate on a pressure range of 70 to 200 kPa (10 to 30 psi) and are an alternative to the high pressure impact sprinklers first used on center pivot and lateral move systems. The half circle device makes it possible to direct the water behind the moving lateral or alter the direction of throw to reduce the application rate.

Spray nozzles produce smaller droplets than those from impact sprinklers. To reduce the tendency to increase wind drift in center pivot and linear move systems the spray nozzles are often mounted on drop tubes, bringing the nozzle down closer to the crop. The throw radius is much less for spray nozzles than for impact sprinklers, particularly if the former are mounted on drop tubes. As a result, application rates are much higher with spray nozzles than with the higher pressure impact sprinklers and application rates may exceed the intake rate of many soils. To help overcome this problem and reduce runoff, the spray nozzles may be mounted on spray booms that wet a larger area at lower intensity.

On rolling terrain, the changes in elevation produce pressure differences along the lateral. These pressure differences may not be significant with high pressure systems. With a low-pressure system, however, the elevation differences may cause pressure changes that are a large percent of the operating pressure, causing unsatisfactory nonuniformity. To compensate for these elevation produced pressure differentials, a pressure regulator is often installed at the base of each spray nozzle of center pivots and linear moves.

New "rotator" sprinkler designs are beginning to replace standard spray nozzles. These sprinkler designs use a special mechanism to produce a slow rotation of the stream without an impact arm. The rotator designs typically provide a much greater radius of throw than do standard spray nozzles, which gives lower instantaneous application rates and better overlap uniformities. They come in a variety of configurations and nozzle angles.

## Undertree Sprinklers

Rotating head low angle undertree impact or geared sprinklers and rotator sprinklers are designed to keep the water jet below the fruit and foliage in orchards. They operate on pressures varying from 70 to 345 kPa (10 to 50 psi) and cover a diameter of 12 to 28 m (40 to 90 ft). When spaced closely (one per tree), they depend on overlap to obtain uniformity. However, the overlapped pattern may be badly distorted by interfering trees unless the jet is below the foliage or the trees are widely spaced. On very wide spacings (1 sprinkler/four trees), they may not overlap.

Small spinners and low pressure spray nozzles are designed to wet the area between trees, cover a diameter of 5 to 12 m (15 to 40 ft), and operate in the low pressure range. Sprinkler discharge ranges from 30 to 570 lph (8 to 150 gph). They are not usually designed to overlap.

## Perforated Pipe

Perforated pipe sprays water from small holes, 1.5 mm (1/16 in) diameter or less. The holes are spaced in a pattern of several rows along the pipe to provide width and uniformity with overlap of only 1 to 2 m (3 to 6 ft). They cover a width of from 3 to 15 m (10 to 50 ft) in response to pressure changes from 30 to 130 kPa (4 to 20 psi). At these low pressures, the jets do not break up into a spray. These conditions result in a high minimum application rate of 12 mm/hr (0.5 in/hr) and limit the use to high intake rate soils. Nozzle lines having a simple row of small outlets, which may form jets or sprays depending on pressure, may be fixed, manually rotated, or mechanically oscillated. They may have very low application rates with oscillating jets and high rates with spray nozzles. They are used primarily for specialty crops and highway landscaping. They constitute a very limited percentage of sprinkler irrigated acreage, but are mentioned here because they do exist.

## TYPES OF SPRINKLER SYSTEMS

Sprinkler irrigation systems use one or more of the sprinkler devices previously described. Water is delivered to the sprinkler by a piping system which may be hand portable, power moved, self-propelled or permanently installed. Numerous ways to classify sprinkler systems have been developed. One method is to group the systems according to the method used to provide coverage to the entire field.

### Hand Move Portable / Lateral Move Portable

A hand move portable system consists of one or more laterals. Laterals are sections of pipeline with sprinkler heads installed at regular intervals along the pipe. Perforated sprinkler lines may function as both lateral and sprinkler. The lateral pipe is generally of aluminum with quick coupling connections at each pipe joint. Pipe section lengths are generally 6, 9, or 12 m (20, 30, or 40 ft). The sprinkler lateral is "set" in one location until the desired amount of water has been applied. he lateral line is then disassembled and carried to the next "set". When the lateral line set has been moved completely across the field, it is disassembled and moved back to the starting location. When the supply main is placed in the middle of the field, one or more half laterals may be set on each side and rotated around the field. Since only half of the water supply goes through each section of the mainline, locating the laterals at opposite ends of the field in this layout makes it possible to reduce the size of the main delivery line. Several half laterals may be used on long fields. Typical hand move lateral systems are shown in Figures 5-1 and 5-2.

Rotating head impact sprinklers are generally used for agricultural crop irrigation, but newer non-impact plastic sprinklers of the rotator design are gaining popularity, especially when the pipe is configured in a "solid set" design for high value produce crops. The sprinkler head is installed on a pipe riser so that it operates above the crop being grown. The risers may be installed in the pipe coupling, but for ease in moving, are usually installed in the center of a pipe length. The length of pipe joint is selected to correspond with the desired sprinkler spacing. This system has a low initial cost but has a high labor

Figure 5-1. Typical Hand Move Lateral Sprinkler System.

Figure 5-2. Hand Move Lateral Sprinkler System. Laterals are Approximately 200 Meters Apart.

requirement for carrying out the irrigation. This type of system can be used on all types of topography and on most crops. With some crops, such as field corn, the difficulty of moving the lateral increases to an impossibility as the crop reaches maturity. On bare sticky soils, moving becomes very difficult because the irrigator must walk through the wet soil.

The biggest single improvement an operator can make to improve the DU is to use alternate sets. With alternate sets, on every other cycle of irrigation, the pipe is placed in an intermediate position. Over the course of 2 irrigation cycles, this effectively cuts the spacing of the laterals in half. There is no increase in labor to use alternate sets.

**End-tow Lateral**

The end-tow system is designed and operated similar to the hand move system. The main difference is in the method of moving the lateral from one set to another. The pipe joints are connected with semipermanent couplers so they can be towed endwise without becoming disconnected. The lateral line is mounted on skids or small wheels that are aligned parallel to the lateral line. After a set is completed, the half field width laterals are towed across the central supply line to the opposite side of the field while being zigzagged ahead. Subsequent sets are made by towing the line back and forth across the supply line. A strip of field up to 70 m (200 ft) wide must be available for the zigzag advancing motion. The use of guide bumpers can reduce this width. This strip is usually seeded to a grass or hay crop. When tall crops, such as corn, are grown, a three or four row width strip must be left bare or planted to low growing crops at each "set" location. A typical end-tow lateral system is shown in Figure 5-3.

Figure 5-3.    Typical Towable Lateral Sprinkler System.

The end-tow system is best suited to square or rectangular shaped fields with fairly uniform topography. It is better suited to close growing crops than for row-crop. Traditional end-tow systems are slightly higher in initial cost than hand move systems because of the greater strength required in the lateral pipe and couplings. They have the lowest cost of the mechanical move sprinkler systems, but require use of a tractor. Care must be exercised in moving the lateral to prevent damage to the pipe and risers and keep maintenance costs at a minimum. Some newer designs use plastic pipe, and are configured as solid set systems but can be removed by pulling them from the end of the field.

## **Side Roll / Wheel Line**

The side roll lateral system is a variation of the hand move lateral sprinklers previously discussed. Irrigation is accomplished in the same moving sequence as with the hand move system. The lateral line is mounted on wheels with the pipe forming the axle. Each joint of torque-tube aluminum pipe has a sprinkler and a wheel. The wheel height is selected depending on the crops that are to be irrigated. The wheels must be of sufficient height to support the lateral line above the crop being irrigated.

A few systems exist in which there is an additional sprinkler at the end of a "tag line", for each pipe joint. As the torque tube rotates, a special rotating coupler and gasket allow this tag line to be dragged behind the unit. This configuration enables the machine to irrigate twice the area per set as with a normal wheel line.

Wheels may sink down in some soils when wet. Large crops also exert a large friction on the wheels as they rotate, so a larger diameter torque tube is often used when there will be appreciable drag. Practical heights of the torque tube are limited to a maximum of abut 1.0 meter.

A drive unit, most commonly powered by an air-cooled gasoline engine located near the center of the lateral, is used to moved the system from one set to another. An operator shuts off the water, disconnects the water supply, walks to the center of the unit while the unit is draining through automatic drain valves below each sprinkler, and starts the engine. Once the lateral is in the new location, the water supply is reconnected and the sprinklers begin to operate again.

Set widths must be in multiples of the wheel circumference and, if self-erecting risers are not used, positioned with the sprinklers in an upright position. Self-erecting risers allow some flexibility in positioning, but require that the lateral be positioned within plus or minus 90° of a vertical sprinkler riser position. Sprinkler heads are normally spaced midway between the wheels. Connection of the lateral to the main line pipe is usually made with quick-coupling aluminum pipe or a short section of flexible hose. A typical side roll lateral system is shown in Figure 5-4. As with hand move sprinklers, positioning the sprinklers in an intermediate position on every other irrigation cycle can drastically improve the overlap uniformity.

Side roll lateral systems are particularly sensitive to damage from high winds. If high winds are expected, the wheels need to be physically staked to the ground and the sprinklers should be operated to fill the pipes with water. If this is not done, the wind can push these machines several kilometers away.

**Side Move Lateral**

The side move lateral system is moved across the field similar to the side roll system. It differs from the side roll system in that the pipe lateral is supported by a structural frame on which the wheels are mounted. The supply pipe does not rotate, as happens with side roll systems. A drive shaft supplies power to each set of wheels for moving the system from one set to another. Water is supplied to the lateral in the same manner as with the side roll system. In order to increase the area covered in one set, trailing lines containing up to 10 sprinklers may be used on the back side of the lateral. Set widths up to 100 m (300 ft) can be made with these systems. The width of the set will be determined by the number of sprinklers used on the trailing lines. Because the sprinklers can be easily spaced close together, higher uniformity and less evaporation may result compared to single line sets. Returning back to the original starting location requires labor to disconnect and reconnect the trailing tubes and re-align the lateral. A typical side move lateral system is shown in Figure 5-5. These have almost universally been replaced by linear move systems, because of the simplicity and mechanical advantages of linear moves. Side move laterals are included in this section only because they are still discussed in other literature.

Figure 5-4.    Typical Side Roll Lateral Sprinkler System.

Figure 5-5. Typical Side Move Lateral Sprinkler System. (Such Large Units Are No Longer Manufactured.)

Both the side roll and the side move system are best suited to fields having a regular shape and relatively gentle topography. They are also best suited to close-growing crops and low-growing row crops. The labor requirement for wheel-move is about half that required by the hand move system. One disadvantage of these systems is that after a field has been irrigated, the system must be moved back to the original side of the field to begin the next irrigation. Again, with the newer equipment available for linear move irrigation, this method of irrigation is now primarily only of historical interest.

## Traveling Gun and Rotating Boom Systems

Traveling gun and rotating boom systems are high volume, high pressure systems. With both of these systems, the application rate is determined by the sprinkler design, water pressure and rate of advance. The amount of water applied at one irrigation is determined by the travel speed of the system. Both of these systems may be operated in a stationary position for a desired time and then moved to a new location. Their most common use is as a continuous move system.

The traveling gun consists of a high volume, high pressure sprinkler head mounted on a trailer or skid. Water is supplied through a flexible hose from the main supply line. In some instances where water is supplied through an open ditch, the pump and power unit may also be mounted on the trailer. The trailer is usually moved through the field by a powered winch and cable arrangement. When polyethylene hose is used, the hose is reeled onto a large drum, dragging the sprinkler cart through the field. The cable winch or hose reel is operated either by a water driven mechanism or by an air-cooled engine. A travel lane is required for the trailer and the flexible supply hose. The pressure required to overcome the friction in the flexible supply hose is significant. The traveling gun is usually equipped with a part-circle sprinkler, covering 75 to 80 percent of the circle. This provides a more uniform application of water and leaves a dry strip ahead of the traveling gun onto which the machine may move. Its pattern is easily distorted by wind. A typical traveling gun sprinkler is shown in Figure 5-6.

The boom sprinkler consists of pipe arms that rotate about a center support system mounted on a 4-wheel trailer. Jet action from the nozzles mounted in the arms rotates the boom. Booms are available with diameters of coverage from about 30 to 75 m (100 to 250 ft). The mechanical part of a boom system is subject to wind problems, the pattern only moderately so. A typical boom sprinkler is shown in Figure 5-7. Both of these systems can be used on most crops. Because of the large droplet size and high application rates, they are best suited to coarse soils having high intake rates and to crops providing good ground cover. They seldom have high uniformity, but are well adapted to supplemental irrigation and tall crops.

**Center Pivot Systems**

The center pivot system is a self propelled system with the lateral supported from wheeled towers spaced from 30 to 50 m (100 to 170 ft) apart. The towers are self-propelled so that the sprinkler lateral rotates around an anchor, or pivot point, in the center of the irrigated area. Water is supplied to the lateral through the pivot point. The speed of travel is determined by controls on the tower which is the greatest distance from the pivot point. Alignment controls cause the other towers to move or stop to maintain proper system alignment. The towers have been propelled by water-operated cylinders, revolving jets, electric motors, compressed air, or hydraulic motors. Electric motors are by far most common, with some usage of hydraulic motors.

Generally, the speed of rotation can be varied from 12 to 120 hours per revolution for a typical 400 m (1300 ft) long lateral. The rate of water supply is constant depending on the pump rate and the design of the system, but the application rate increases with an increase in the length of the lateral. The longer the lateral, the faster the end travels and the larger area the end irrigates with each rotation. Consequently, the water application rate required at the end to cover the area during one rotation is higher for longer laterals. If the system is not properly designed and operated, the high application rate at the outer end of the lateral often causes runoff to occur. The depth of water applied during one irrigation depends on the speed of rotation of the lateral. The most common operation is to apply 12 to 25 mm (.5 to 1.0 in) with a 1 1/2 to 3 day rotation cycle. Since the

Figure 5-6.   Typical Traveling Gun Sprinkler.

Figure 5-7.   Typical Rotating Boom Sprinkler.

lateral moves in a circle, the corners of the field are left unirrigated. Special cornering equipment is available to reduce the area not irrigated to 3 to 4 hectares out of 65 hectares (8 to 10 acres out of 160 acres). Other options for the corners include installing a solid set sprinkler system, or using drip irrigation.

Center pivot systems are suitable for irrigating most field crops and have been adapted to irrigate vineyards and dwarf orchards. Since drive towers are spaced 30 to 60 m (90 to 200 ft) apart, the field must be relatively free of obstructions (i.e., buildings, power lines, and other obstructions). Pivot lengths vary from a hundred meters (300 feet) to over 800 meters (2600 feet), with about 400 meters being a common length.

Many innovations, such as narrow bridges across gullies and special openings in fences have been developed to help overcome some of the obstacle limitations. These systems have a moderate initial cost but have a low operating labor requirement. They do require periodic maintenance and good water filtration, but beyond that they can be very easy to operate and to automate.

Soil-bearing strength under irrigation conditions must be adequate to support the weight of the system wheels. Special tires and treads have been designed to make these machines adaptable to a wide range of soils. Sprinkler packages are commonly designed to irrigate behind the wheels rather than on the wheels, thereby keeping the wheel tracks dry during movement.

The trend of pivot sprinklers is definitely toward low pressure sprinklers suspended from weighted drops, with a pressure regulator at each sprinkler. A further description of these sprinklers is in the next linear move section. A typical center pivot system is shown in Figure 5-8.

Figure 5-8.  Typical Center Pivot Sprinkler System. Photo Courtesy of Valmont Industries, Valley, NE.

## Linear Move (Lateral Move) Systems

The linear move system is a self-propelled lateral line that utilizes a tower support system and guidance controls similar to the center pivot sprinkler system. Water delivery to the constantly moving lateral is accomplished similar to the traveling gun or boom sprinklers, utilizing flexible hose or open ditch pickup. A variation with automatic mechanical connection to periodically spaced risers on a buried mainline is also available. The system is designed to operate on rectangular shaped fields that are free of obstructions. Its main advantage is that it covers all the field, has a high application uniformity coefficient, and has a lower application rate than is found on the outer ends of center pivots. Many types of sprinklers or nozzles may be used over a wide range of pressures. The industry has trended toward the use of drop tubes and Low Energy Precision Application (LEPA) or Low Elevation Spray Application (LESA) sprinklers. Both are common on linear move and center pivot systems. A linear move system with sprinklers on booms is shown in Figure 5-9. Linear move sprinkler systems with LEPA/LESA sprinklers are shown in Figures 5-10 and 5-11.

LEPA systems employ drop tubes and very low pressure nozzles (each with a pressure regulator) placed on or close to the field surface to distribute water into blocked furrows. Water distribution is the same as for conventional systems, but the water is applied at the soil surface, reducing evaporation losses common to sprinkler systems. High efficiency irrigation requires either very high soil intake rates or adequate storage in the furrow micro basins to prevent both runoff and non-uniformity along a furrow.

Figure 5-9.   Typical Linear Move Sprinkler System with Sprinklers on Booms. (Since the Early 1990s, Most Systems No Longer Have Booms.)

LESA systems employ drop tubes, but use sprayers located near the ground. This system also requires distributing water into blocked furrow segments because of high application rates. LESA systems spray water onto the plant canopy, whereas the LEPA systems apply water directly to the ground surface. LESA systems may not be applicable with closely spaced crops tall because of the tendency for the sprayers to become entangled in the crop canopy. However, they are excellent for crops such as onions and potatoes.

Figure 5-10.  Typical Linear Move Spinkler System with LEPA Sprinklers. Photo Courtesy of Valmont Industries, Valley,NE.

Figure 5-11.  Typical Linear Move Sprinkler System with LEPA Sprinklers on Onions.

**Solid Set Systems**

A solid set system is a system with a main line and laterals that remain in place during all or part of the growing season of the crop. It requires enough mains and lateral pipes to cover the entire irrigated area. If the main and laterals are buried and left in place from one season to the next, the system is referred to as a permanent solid set system. Where the system is installed on the surface, the system is referred to as a portable solid set system. Portable solid set systems are usually installed after the crop is planted and then removed just before harvest. A typical portable aluminum solid set sprinkler system is shown in Figure 5-12.

Portable solid set systems are also used to germinate seed and establish a crop stand with other methods of irrigation being used after the crop is established. After germination is complete, the portable system may be moved to another field and the field is irrigated by surface irrigation or by drip irrigation.

Permanently installing the system, by burying the main and lateral lines under ground, facilitates farming operations while protecting the system from damage by farm equipment.

The portable solid set system utilizes available labor at the beginning and end of the growing season but minimizes labor requirements during the irrigation season. New plastic lateral pipeline materials and retrieval equipment are beginning to reduce the cost of system retrieval. This new retrieval equipment

Figure 5-12. Portable Aluminum Solid Set Sprinkler System on Carrots.

typically remains on the outer edge of the field, so it is unnecessary to enter the field to remove the equipment.

The solid set system is well suited to irrigating crops that respond to light, frequent irrigations and for use in climate modification. Where the water supply is adequate for large area coverage, crops can be protected from frost and severe high temperatures by continuous, or frequent, low application rate irrigation. However, such application applies water far in excess of evapotranspiration demands and may cause excessively wet soils or elevated water tables. Part-circle sprinklers are often used near the field boundaries to improve the application uniformity on the periphery of the field. The edge sprinklers can be tipped to achieve the same result. Solid set systems have a lower evaporation loss in the air than the single line systems.

The operation of a solid set sprinkler system is sometimes modified by the use of sequencing valves. In such a system, selected sprinklers in each lateral may be operated. For example, the first sprinkler on each lateral may function for the desired period. These are then shut off and the second sprinkler on each lateral functions. In this system, a sequencing valve is installed on each sprinkler. The valves are activated through water pressure change or electrical or air controls connected to automatic controllers. The controls can also be activated through the use of moisture tensiometers installed at key locations in the field. In the sequencing system, the water supply is divided into numerous laterals, permitting the use of much smaller pipe for the laterals.

Solid set systems have high initial investment costs but require very little irrigation labor. When used on field crops, the sprinkler risers are obstacles to normal farming operations. Because of their fixed spacing, they cannot take advantage of the Distribution Uniformity improvement made possible by alternate positioning setting of the lines, as can be done with hand move and side roll sprinkler systems.

## **Undertree Orchard Sprinkler Systems**

These systems may be permanent or portable and may include full overlap or may irrigate one to four trees per sprinkler without overlap. Either low angle rotating head sprinklers or small spinner sprinklers may be used. Usually, the spinner type sprinklers are designed for use without overlap and may be either placed permanently or used portably on movable hoses. The permanent installations can be used at normal frequencies or can be automated and used at frequent intervals with a small soil moisture deficiency similar to drip/micro irrigation operations. Because they wet more of the total soil area, they can operate at a lesser frequency than the drip or micro irrigation methods. A typical undertree orchard sprinkler system is shown in Figure 5-13.

The permanent sprinklers are often spaced on buried laterals in alternate rows and between alternate trees in orchards. When using larger rotating head, low trajectory sprinklers, the system is best adapted to widely spaced trees that can be pruned above the jet. Otherwise, tree interference will badly distort the pattern. Also, wetting of some fruits or tree trunks on certain rootstocks may

Figure 5-13   Typical Undertree Orchard Sprinkler System.

cause damage or enhance disease problems. Stream splitters are available for some sprinkler models to eliminate water on the trunks. Saline water on leaves may be absorbed and cause leaf drop. When smaller spinner type sprinklers are used, closer tree spacing without interference is possible. In all cases, low trajectory angle sprinklers should be used to reduce foliage wetting and canopy interference. Cultural practices must be considered in placing the sprinklers.

Portable sprinkler systems called "hose-pull" are sometimes used, but in the U.S. have mostly been replaced by drip/micro irrigation systems. Hose-pull systems usually have 2 to 5 sprinklers installed in series on a small diameter flexible PVC irrigation hose that is pulled up and down several rows (usually 2 to 4) while attached to a single outlet. Alternate side sequencing is normally employed. Portable pipe can be used in place of the hose-pull arrangement, but the pipes are difficult to move in mature orchards and their use is not common.

Overlapping sprinkler patterns plus possible tree interference may result in excessively high and low areas of application which may be detrimentally related to an individual tree. The depth of application in the overlap area should be considered in design and management. Excessive deep percolation may occur in these areas if the application is far in excess of the average application. If the variation in application within the root area of one tree is not excessive and is consistent with time, the extraction pattern may adjust to accommodate the variation. If the variation in application depth occurs from tree to tree, extraction pattern adjustment will not alleviate the non-uniformity and lower efficiency will result.

Non-overlapping sprinklers of any type, covering nearly all of the area between four trees, or just an individual tree, can result in very high potential irrigation efficiency as deep percolation losses can be very small. Since the application pattern of these sprinklers decreases to zero at the periphery, an appreciable area of under-irrigation, from which no deep percolation occurs, results. With proper selection of sprinklers or adjustments, the small over-irrigated area can have a very uniform depth and be controlled to have very little deep percolation loss. Since the entire soil mass is not wetted, frequency must be increased over that of a completely wetting system. Undertree sprinkler systems can be very simple to operate and maintain.

## **Overtree Orchard Sprinkler Systems**

High risers above the tops of the trees permit the use of normal open field sprinklers. High trajectory heads should be used so that the drops are falling nearly vertically to alleviate the tree interference problems that occur using low trajectory nozzles. The relative height of the riser and the tree changes as the trees grow and riser height may also need to be varied.

This system can be used for climate control in orchards and vineyards. It is practical to cool the plants to delay an abnormally early bloom that might subsequently be subject to frost. Also, for excessively hot periods, cooling may be practical. During periods of frost, prevention of frost damage may be obtained by sprinkling to provide water which will freeze on the plant, providing an ice coating that prevents the plant's temperature from becoming detrimentally low. The development of an ice load may cause limb breakage and additional water use may cause a detrimentally wet soil or high water table.

## CAPABILITIES AND LIMITATIONS

There are so many different types of sprinkler systems that it is difficult to generalize about advantages and disadvantages. However, some major advantages are:

1. Irrigation scheduling is relatively simple, as the system dictates the application rate, which is predictable.

2. Because the sprinkler spacing, nozzle size, and pressure are typically fixed by the design, there are very few management options. This means that the management emphasis is more on maintenance and scheduling than on strategies for obtaining a good DU (such as with surface irrigation).

3. Land grading requirements are minimal. This is extremely important in many developing countries where the fields are small and good land grading is almost impossible to achieve.

4. Labor requirements with some of the sprinkler irrigation systems, in particular center pivots, is very small.

Some key disadvantages of sprinklers as are:

1. They may require more pumping energy than surface and drip/micro irrigation methods.

2. Sprinkler methods require better source water filtration than do the surface irrigation methods (but less than drip/micro).

3. Some of the sprinkler methods (especially hand move) are labor intensive in terms of physical exertion required. For this reason, it is sometimes difficult to find irrigators.

4. Sprinkler methods that apply water to leaves are unsuitable for irrigating foliage if the water is salty.

## Crops

Nearly all crops can be irrigated with sprinklers. The characteristics of the crop, especially the height, must be considered in selecting the type of system. For example, a side roll sprinkler system is not feasible in irrigating field corn and other tall crops. There are also some special crops for which sprinklers are not desirable; for example, onions for dehydration may discolor if their tops are wet with sprinklers, rot may increase on tomatoes, and timothy hay lodges easily if sprinkled.

Sprinklers are sometimes used to germinate seed and establish ground cover for such crops as grass sod and alfalfa. They are commonly used for pre-irrigations on soils with high initial intake rates. Light, frequent applications may not be economical and are more difficult to affect with surface irrigation methods, but can be easily managed under sprinkler systems.

Crop damage due to weather extremes, such as high temperatures or frost, can be reduced. Special designs to facilitate continuous application of water are required.

## Soils

Most soils can be irrigated by some sprinkler method. However, sprinklers are not easily suited to soils having less than 3 mm/hr (0.12 in/hr) final intake rate unless special measures are used to increase intake or provide uniform surface ponding to control runoff.

To allow for pattern non-uniformity, soils should have a final intake rate for the irrigation period greater than 1.3 times the average application rate of the sprinkler. Special land treatment, such as surface mulching or shaping of the soil surface, can permit the use of sprinklers on soils where the intake rate is too low to match the sprinkler application rate. Surface ponding by such techniques as contour farming or the use of furrow dams can aid in adapting sprinklers with a high application rate to lower intake rate soils.

## Topography

Topography can vary from flat to fairly steep and rolling and still be suitable for sprinkler irrigation by one of the several types of systems that are available. Land leveling is not normally required. Some surface grading or smoothing may be required where surface drainage of rainfall is needed. In some instances, surface grading may be beneficial to permit earlier spring farming operations and to aid in harvesting by reducing or eliminating wet areas.

In general, sprinkler irrigation can be used on any topography that is suitable for farming but may be restricted by cultural practices, gullies, poorly drained spots, etc. The hand move and solid set systems are well adapted to steep and rolling terrain. Of the mechanical move systems, only center pivot systems are suitable for steep and heavily rolling terrain. The other mechanical move systems are more limited. Flow control devices and pressure regulators for individual sprinklers are widely available.

Sprinklers are applicable to soils on rolling topography or that are too shallow to permit surface shaping or too variable to permit efficient use of surface irrigation methods.

## Water Supply

Sprinkler irrigation systems require an essentially constant rate water supply that is available 24 hours per day for the least costly, most energy efficient designs. Water supplies that are available on a rotation basis require excess system capacity or on-farm storage, both of which add to the cost of the system. Although the water supply should be available 24 hours per day, it must be flexible in duration so that set times can be adjusted to the SMD.

## Salinity/Water Quality

Salinity can be a factor in selecting the irrigation method. Salts can be leached from the soil by sprinklers with high uniformity. Less water is required than with flooding methods as water moves through smaller soil pores in an unsaturated condition. This is important in areas with a high water table. The method used to leach salts is influenced by the soil and topography.

Many crops are sensitive to foliar absorption of salts dissolved in the irrigation water. Defoliation may take place where the water strikes the leaves. Under low humidity conditions, water on the leaves and salts become more concentrated due to drying between sprinkler rotations. Irrigating at night can help alleviate this problem. It is also recommended that sprinkler heads that rotate at one revolution per minute or faster be used under such conditions. Very low application rates should be avoided.

Surface water supplies typically require some filtration to remove debris that could plug orifices. The lower the sprinkler discharge, the finer the filtration requirement. To reduce nozzle wear and lateral sedimentation, sediment-laden waters require settling before entering the system.

## Climate

Climate is a factor in the selection of the method of irrigation. In the higher rainfall, higher humidity areas, sprinklers are well suited to the supplemental application of water for crop use. While very low humidity has an effect on sprinklers due to direct evaporation of the spray before the water hits the soil, the most significant climate factor in selecting the method of irrigation and the type of sprinkler system to be used is wind. High pressure guns and booms that are designed for wide area coverage are not recommended in areas where strong winds prevail. Under these conditions severe distortion of the application pattern will occur. The use of LEPA and LESA sprinklers on center pivots and linear-moves virtually eliminates sensitivity to wind.

## **Efficiency**

Efficiencies that can be attained with sprinklers (Table 5-1) depend on a number of factors. In practice, the efficiencies range from very low to very high depending on the system selected, the design of the system and its operation. It is fairly easy to schedule irrigations with almost any type of sprinkler irrigation method, with scheduling with center pivots, linear moves, and solid sets being the easiest. Evaporation losses vary, but can be in the 1 - 20% range, depending upon the nozzle type, height of trajectory, climate conditions, extent of wetting of the crop canopy, and numerous other factors. A well designed and properly operated sprinkler irrigation system will have little or no runoff.

Therefore, the Distribution Uniformity of the sprinkler systems becomes a key element in determining the potential efficiency of operation. The fixed grid sprinkler systems that depend upon overlap in 2 dimensions and which cannot be moved (solid set systems on field crops) typically have the lowest DU. The main factors in system DU for these systems are (1) flow rate uniformity, which is influenced by pressure differences, nozzle size variations, and plugging and wear, and (2) overlap uniformity. The poor overlap uniformity dominates the overall system DU for these systems.

Table 5-1.   Attainable Application Efficiencies ($AE_{lq}$) for Sprinkler Irrigation Systems.

| System Type | Attainable Application Efficiency |
|---|---|
| Hand Move, End-tow side roll laterals (highest w/ alternate sets) | 65 - 85% |
| Traveling gun, boom | 60 - 75% |
| Center pivot, linear move | 75 - 90% |
| Solid set, side move | 70 - 80% |
| LEPA | 80 - 93% |
| Undertree orchard (non-overlap) | 80 - 93% |

The next highest sprinkler system DU values occur with sprinklers on a grid (hand move, side roll, lateral) that can be moved. By using alternate sets, the overlap uniformity can be appreciably improved over the course of 2 irrigations.

The next best DUs will occur with well designed and maintained linear move and center pivot systems. Assuming that the movement of the machines is smooth (which is possible over the course of several passes), overlap non-uniformity only occurs in one direction -- between the sprinklers themselves. In the second direction (the direction of travel), the continuous movement eliminates non-uniformity. Because sprinklers can be placed close to each other, and because pressure regulators are commonly used, such irrigation methods can have $DU_{lq}$ values above 0.90.

The highest DU values occur on permanent, solid set sprinkler systems on orchards. If those sprinklers are on a close spacing (one sprinkler per tree or per 2 trees), the overlap pattern does not impact the uniformity of water between plants. Therefore, the DU is only influenced by the flow rate non-uniformity. Because these are permanent systems it is relatively easy to filter the water to eliminate plugging and wear, and pressure differences can be controlled in the initial design procedures. It is common to have $DU_{lq}$ values for well managed and designed undertree sprinkler systems in the 0.93 - 0.95 range.

## Irrigation Scheduling

Irrigation scheduling is relatively simple with sprinklers, but it has fundamental differences when contrasted to drip/micro and surface irrigation methods. Even within the various sprinkler methods there are fundamental differences.

Hand move and side roll sprinklers are typically irrigated for a known duration of 12 or 24 hours (less the move time). Therefore, the application depth is determined in advance by that set duration. Good scheduling involves waiting until the SMD matches the actual application depth (adjusted for Application Efficiency). This contrasts with drip/micro for which the application depth and time are adjusted to match the SMD.

Center pivots and linear-moves have a fixed application rate but have variable speeds (which translates into a variable duration). They require some minimum amount of time to complete a cycle. Scheduling typically involves a computation of the necessary hours/cycle (rotation) or hours/week to apply the desired depth. With moving systems such as center pivots and linear-moves, irrigation schedulers must balance the benefits of small SMDs - which tend to increase the percentage evaporation losses from wet foliage, and larger SMDs - which reduce the percentage evaporation loss but increase the potential for runoff.

# INSTITUTIONAL CONSIDERATIONS

## Labor

Labor requirements of sprinkler systems vary depending on the level of automation and mechanization of the equipment. Hand move systems require the least skill but greatest amount of labor. Side roll and laterals require somewhat less labor, but a higher skill level. Portable solid set systems require less labor but the same skill level as hand move systems. Considerable skill is required to operate center pivot, linear move and LEPA systems, but the amount of labor is very low. In fact, management and maintenance labor are the main requirements. The least labor is required by permanent solid set systems. Table 5-2 presents typical labor requirements for various sprinkler irrigation methods.

## Service Availability

Outside service requirements vary depending on the degree of sophistication of the equipment. All systems require maintenance capability, either on-farm or outside, to repair pipe. Electrical and mechanical service may be required on more mechanized equipment such as center pivot and linear move systems. This service can be provided from outside service organizations or by training of farm personnel.

Table 5-2. Operating Labor Requirements for Sprinkler Irrigation Systems.

| System Type | Operating Labor | |
|---|---|---|
| | man-hrs/ha-mm | man-hrs/ac-in |
| Hand move lateral | .017 | .175 |
| End tow lateral | .010 | .103 |
| Side roll lateral | .012 | .123 |
| lateral | .010 | .103 |
| Traveling gun | .007 | .072 |
| Center Pivot | .001 | .010 |
| Center Pivot w/corner | .001 | .010 |
| Linear move w/ditch | .002 | .021 |
| Linear move w/pipe | .002 | .021 |
| Portable solid set | .010 | .103 |
| Permanent solid set | .001 | .010 |

# ECONOMIC FACTORS

## Capital Cost

Table 5-3 shows capital cost ranges in terms of 1995 U.S. dollars for the various sprinkler irrigation methods. These tables appear with the sections describing the methods.

## Energy Cost

Energy requirements vary widely among system types. Unless the water source is elevated sufficiently above the land surface to provide gravity pressure, some energy source for pump operation is needed for all systems. At the low end are the LEPA systems, requiring 100 to 140 kPa (15 to 20 psi). Center pivot and linear move systems require from 170 to 550 kPa (25 to 80 psi) depending on sprinkler design. Set systems require from 170 to 330 kPa (25 to 50 psi). Traveling guns require from 690 to 825 kPa (100 to 140 psi). The energy source may be electric, gasoline, diesel, liquid propane or natural gas. Solar energy or wind energy is being used on an experimental basis, but is not yet cost effective.

Energy costs are so regional in nature that actual costs cannot be provided. The energy requirements shown in Table 5-4 may be used to estimate the energy cost utilizing the appropriate unit cost. All figures are based on a pump efficiency of 75% and references to hectare millimeters of water are on a gross application basis for pressure indicated under each method.

Table 5-3.  Capital Cost Ranges for Sprinkler Irrigation Systems.

|  | Area Irrigated | | Capital Cost Range | |
|---|---|---|---|---|
| System Type | ha | ac | $/ha | $/ac |
| Hand move lateral | 65 | 160 | 450-675 | 175-275 |
| End tow lateral | 65 | 160 | 600-950 | 250-400 |
| Side roll lateral | 65 | 160 | 800-1100 | 325-450 |
| lateral | 50 | 120 | 950-1350 | 400-550 |
| Traveling gun | 30 | 80 | 950-1200 | 400-500 |
| Center Pivot | 80-50 | 200-125 | 700-1100 | 275-450 |
| Center Pivot w/corner | 60 | 150 | 950-1200 | 400-500 |
| Linear move w/ditch | 130 | 320 | 1100-1300 | 450-525 |
| Linear move w/pipe | 130 | 320 | 1600-2050 | 650-825 |
| Portable solid set | 65 | 160 | 2700-3250 | 1100-1300 |
| Permanent solid set | 65 | 160 | 2300-3500 | 925-1400 |

Table 5-4. Energy Requirements for Sprinkler Irrigation Systems.

| System Type | Energy Required | |
|---|---|---|
| | kW-h/ha-mm | kW-h/ac-in |
| Hand move lateral | 0.9-2.1 | 9.2-21.6 |
| End tow lateral | 0.9-2.1 | 9.2-21.6 |
| Side roll lateral | 0.9-2.1 | 9.2-21.6 |
| lateral | 1.1-2.1 | 11.3-21.6 |
| Traveling gun | 3.5-4.9 | 36.0-50.4 |
| Center Pivot | 0.9-2.3 | 9.2-23.7 |
| Center Pivot w/corner | 1.0-2.4 | 10.3-24.7 |
| Linear move w/ditch | 0.9-2.3 | 9.2-23.7 |
| Linear move w/pipe | 1.2-2.6 | 12.3-26.7 |
| Portable solid set | 0.9-2.1 | 9.2-21.6 |
| Permanent solid set | 0.9-2.1 | 9.2-21.6 |

## Labor Cost

Operating labor requirements vary by system type. The labor requirements of Table 5-2 can be used to estimate labor costs for the different system types. The requirements are listed as person-hours per hectare-millimeter of gross application.

## Management Cost

Management time has not been estimated separately. Utilizing the above operating labor costs will cover most management cost with the exception of decisions on when to irrigate and how much to apply. A typical cost for this level of management is generally $12 to $25 per hectare ($5 to $10 per acre) per crop season, depending on the length of season and intensity of irrigation management required.

## Operation and Maintenance Costs

Operation costs are included in the energy and labor requirements. Average annual maintenance costs may be estimated by applying percentages of initial capital cost.

Maintenance cost for LEPA systems are the same as linear move systems. Maintenance costs are based on the equipment lives, shown in Table 5-5, and include maintenance of pumps and mainlines.

Table 5-5. Average Annual Maintenance Costs for Sprinkler Irrigation Systems.

| System Type | Life-yrs | Maintenance, % of capital cost |
|---|---|---|
| Hand move lateral | 15 | 2 |
| End tow lateral | 10 | 3 |
| Side roll lateral | 15 | 2 |
| lateral | 10 | 6 |
| Traveling gun | 15 | 6 |
| Center Pivot | 15 | 5 |
| Center Pivot w/corner | 15 | 6 |
| Linear move w/ditch | 15 | 6 |
| Linear move w/pipe | 15 | 6 |
| Portable solid set | 15 | 2 |
| Permanent solid set | 20 | 1 |
| Orchard hose-pull | 15 | 3 |

## Special Cultural Costs

Some systems require special cultural costs. Center pivot or linear move systems on low intake soils may require the use of micro-basin tillage equipment at an annual cost of about $17 per hectare ($7 per acre). LEPA systems also require micro-basin equipment. Permanent solid set systems on row crop land will require special cultivation practices. Traveling gun systems may require tow lane maintenance in some crops.

# CHAPTER 6
# SUBIRRIGATION - WATER TABLE MANAGEMENT

## DESCRIPTION

Water table management (subirrigation) is accomplished by controlling the water table, usually through a subsurface drainage system. The crop is subirrigated with water being furnished directly to the root system by capillary action from the saturated portion of the soil profile referred to as the water table or perched water table, or by wetting the soil by quickly raising and lowering the water table. The applicability is normally in humid areas where drainage of the soil profile for adequate root development is necessary or beneficial for crop production or when drainage is economically beneficial from a trafficability standpoint for crop production or harvesting. The water table is controlled at some elevation above the subsurface ditch or pipe drain during the irrigation season by use of one or more water control structures, in a main outlet ditch or collector tube or pipe, which usually employs a weir overflow system or pumps. As water is being utilized by the plants, it must be replenished through the system by a well or some surface irrigation water source.

## CAPABILITIES AND LIMITATIONS

### Crops

Subirrigation systems are adapted to most any medium to shallow rooted crops or orchards. Since the water is supplied and moved to the root zone for plant use from below the ground surface, the irrigation process or system does not interfere with surface activities or cultivation needs of the crop.

### Soils

This type of irrigation system should only be proposed on soils that drain quickly and have a moderately high water table. To avoid subsidence, it has been used on organic soils that oxidize readily when the water table is excessively lowered for appreciable time periods. Raising the water table to get complete saturation of organic soils for part of a dormant season has also been used for nematode control.

Coarse grained soils that are generally associated with a low water holding capacity and have a relatively high hydraulic conductivity in the saturated state are well suited to this type of irrigation system. Because of dual use, it is more economical if soil profile drainage is also needed for parts of the year.

### Topography

Level to mild uniform slopes are desirable to facilitate design, layout, and management of a subirrigation system, but they are not mandatory. The subsurface drainage system should be designed to have a minimum positive slope for drainage. The drain system could be laid almost on the contour while the collector channel or pipe containing control structures would be on the steeper slope.

The shape of the field need not be constrained in any way to accommodate this type of system as it is underground and independent of the cropping or orchard layout.

### Water Supply

The water supply rate is designed to accommodate the evapotranspiration rate plus perimeter and deep seepage losses. Since site adaptability is dependent on maintaining a high water table during at least part of the year, deep seepage losses should not be a major consideration. Perimeter losses are of less and less concern, as the size of field increases. Fields of 100 acres or more are practical. Both deep seepage and perimeter losses can be computed. Since the water is supplied beneath the soil surface, the direct evaporation losses should be minimal.

### Salinity / Water Quality

At this time, water table management systems have not been recommended for use in salinity sensitive areas or to utilize irrigation water containing appreciable salts. These systems are normally installed in humid areas, where leaching water to offset saline buildup is available through natural precipitation. This concern needs to be analyzed on a site specific basis.

### Climate

A subirrigation system is not affected by wind or excessive evaporation conditions as the supply and distribution is below ground. Frost control is not provided, nor is cooling by evaporation. Since these systems are used on naturally wet soils, there is danger of excessive wetness during high precipitation periods. This can be prevented by adequate consideration of drainage requirements during design or appropriate management of water level control weirs as needed.

### Efficiency

Based on current information, subirrigation can be a very efficient irrigation method as far as water conservation is concerned (i.e., 80 percent or better application efficiency on field size units where 8 inches of water pumped may furnish 6 to 7 inches for evapotranspiration). Such a high number, of course, depends upon the existence of a natural high water table and little or no lateral movement of the water outside of the field boundaries.

## INSTITUTIONAL CONSIDERATIONS

### Labor

The labor requirement for subirrigation is very low to nonexistent once the system is set. This system can be automated quite easily. A constant water level, at least for any given portion of a crop season, can be controlled by a float or electrodes that turn an electric pump on or off. If a changed water level is desired, as a crop develops, this can be very easily adjusted by changing weir levels at a convenient time. For some crops or management techniques, there may be no need to change weir setting from the beginning to the end of the irrigation season. An alternate procedure is to quickly (one day) raise the water table to wet the soil from below and then quickly drain it, rather than depending solely on capillary rise.

### Management Skills

As with any irrigation system, there are required management skills. These are mostly related to understanding the soil-water-plant relationship. Monitoring the soil moisture or water table level within the field is necessary for effective management.

### Service Availability

Service facilities are not a concern except for upkeep of the pump and well system. The worst possible condition would be the need to repair or clean a clogged subsurface tubing system, which should be extremely rare, or the maintenance of the combination supply-drain ditches.

## ECONOMIC FACTORS

### Capital Cost

Development costs are primarily related to the costs of land grading and the pipe installation cost of the subsurface drain-supply system, which may be approximately 300 to 750 m/ha (400 to 1,000 ft/ac), depending on the drain spacing. This cost might vary from $500/ha ($200/ac) (minimum length without filter) to possibly $2,500/ha ($1000/ac) (maximum length with a filter). On permeable stable organic soils, open ditches can also be used effectively at a cost of about $250/ha ($100/ac).

In order to assure that the system will maintain the water table at the desired stage at the midpoint between drains, the spacing of subsurface drains may be intentionally reduced for a subirrigation design. A reduction in spacing by as much as 20 percent is reasonable. For average cost purposes, a 20 m (65 ft) spacing which includes a thin spun nylon type of filter for $3.20 per meter would cost $1,900/ha ($770/ac). An outlet ditch with control structures might cost $50/ha ($20/ac). The establishment of a well might be $6,000 or $150/ha ($60/ac) for 40 ha (100 ac) (assumed 100 m well at $60/m). Pump and motor to supply a peak rate of 7.6 ha-mm (0.3 ac-in) per day for 40 ha (100 ac), plus losses, should be perhaps 25 kW (30 h.p.), at an estimated cost of $4,000 or $100/ha ($40/ac).

Total capital costs are estimated at $2,200/ha ($890/ac). Note that the cost of subsurface drains is approximately 85 percent of the total. Wider drain spacings on more permeable soils would be considerably less expensive.

## Energy Cost

The pumping rate needed will not significantly exceed the evapotranspiration rate (perhaps 7.5 mm/day) for a peak period, which is approximately 35 lps (560 gpm) for 40 ha (100 ac). For a lift of 45 m (150 ft), supplying 44 lps (700 gpm), would require 20 kW (26.5 hp). Therefore, a liberal estimate of pumping time for 200 mm (8 in) of irrigation on 40 ha (100 ac) is 516 hours. In areas where the control of a permanent high water table is the basis for the system, very low lift pumping is all that is required with an energy cost of as little as 20% of the above cost.

## Labor Cost

This is a minimal consideration, as it is really a management consideration and for overseeing operation. Inspection of the pumping system, adjusting the weirs and occasionally examining the soil moisture condition in various parts of the field are all that is involved. Refueling would be necessary on a gas or diesel pump motor.

## Management Cost

Management is very important in this type of system but, for a cost evaluation, should be looked at as general farm management overhead.

## Operation and Maintenance Cost

The only significant operating cost is in conjunction with pumping of the irrigation water. As noted above, a maximum of 516 hours of pumping with a 20 kW (26.5 hp) load is approximately 11,700 kWh. At the rate of $.07/kWh, operation costs would be $820 for 40 ha (100 ac) for a season or approximately $20/ha ($8/ac).

Capital or development costs for this type of system should be considered on a long-term basis, at least over 20 years. In conjunction with maintenance or operation of the system, the collector ditch and structures may need a $500 investment approximately every 5 years or $2.50/ha ($1.00/ac) per year. Another $2.50/ha ($1.00/ac) per year should cover pump and well maintenance.

# REFERENCES

ASAE. (1990). Standards 1990. Standards, Engineering Practices and Data. 37th Edition. ASAE. St. Joseph, MI 49085.

Burt, C.M., A.J. Clemmens, T.S. Strelkoff, K.H. Solomon, R.D. Bliesner, L.A. Hardy, T.A. Howell, and D.E. Eisenhauer. (1997). Irrigation Performance Measures: Efficiency and Uniformity. Journal of Irrigation and Drainage Engineering 123(6):423-442.

Burt, C.M. and J. Keller. (1977). Very Low-Pressure Sprinkler Irrigation, NSAE Paper No. 76-2517.

Burt, C.M. and S. W. Styles. (1994). Drip and Microirrigation for Trees, Vines, and Row Crops. Irrigation Training and Research Center (ITRC), Dept. of BioResource and Agricultural Engineering, Cal Poly, San Luis Obispo, CA 93407.

Burt, C.M., R.E. Walker, S.W. Styles, and J. Parrish. (1995). Irrigation System Evaluation Manual - rev. 1995 for Windows. Irrigation Training and Research Center (ITRC), Dept. of BioResource and Agricultural Engineering, Cal Poly, San Luis Obispo, CA 93407.

Cachuma RCD. (1994). Final Report - Irrigation Water Management Program - Santa Barbara and San Luis Obispo Counties. Cachuma Resource Cons. District, Santa Maria, CA 93454.

Clemmens, A.J. and A. Dedrick. (1981). Estimating Distribution Uniformity in Level Basins. Trans. of the ASAE 24(5): 1177-1180, 1187.

Hart, W.E. (1961). Overhead Irrigation Pattern Parameters. Agric. Engr. July. 354-355.

Hart, W.E. and W.N. Reynolds. (1965). Analytical Design of Sprinkle Systems. ASAE Transactions 8(1):83-85,89.

ITRC. 1997-99. Results of On-farm Irrigation Evaluations for Distribution Uniformity. Unpublished results of field evaluations in California and Oregon. Irrigation Training and Research Center. California Polytechnic State University. San Luis Obispo, CA 93407.

Keller, J. and R.D. Bliesner. 1990. Sprinkle and Trickle Irrigation. Van Nostrand Reinhold Publishing. New York. 652 p.

Kemper, W.D., D.C. Kincaid, R.V.Worstell, W.H. Heinemann, T.J. Trout, and J.E. Chapman. (1985). Cablegation Systems for Irrigation: Description, Design, Installation, and Performance. USDA/ARS. ARS-21. 208 p.

Kubota (1986). Kubota Surge Flow System. Cat. No. 971-4563. Information brochure by Kubota America Corp. Los Angeles, CA.

Little, G.E. (no date). Distribution Uniformity Variability of Pressure Irrigation Systems for Orchards. Unpublished M.S. Thesis. Univ. of Calif., Davis.

Merriam, J.L. (1966). A Management Control Concept for Determining the Economical Depth and Frequency of Irrigation. Transactions of the American Society of Agricultural Engineers 9(4):492-498.

Merriam, J.L. and J. Keller. (1978). Farm Irrigation System Evaluation: A Guide for Management. Utah State University, Logan, Utah.

Mission RCD. (1993). A Summary of Agricultural Irrigation System Evaluations in Northern San Diego County - 1983-1992. Mission Resource Cons. District. Fallbrook, CA.

On-Farm Irrigation Committee, Irrigation and Drainage Division, ASCE. (1978). Describing Irrigation Efficiency and Uniformity. Journal of the Irrigation and Drainage Division, ASCE. 104(1): 35-41.

On-Farm Irrigation Committee, Irrigation and Drainage Division, ASCE. (1984). Recommended Irrigation Schedule Terminology. Proceedings of the Specialty Conference, ASCE, July 24-26. Flagstaff, AZ. pp 219-221.

Styles, S. W., and C. M. Burt. (1996). Evaluation of Subsurface Drip Irrigation on Peppers. Cal Poly Irrigation Training and Research Center. Report submitted to the California Energy Commission.

Replogle, J.A., J.L. Merriam, L.R. Swarner, and J.T. Phelan (1981). Farm Water Delivery Systems. Chapter 9 of Design and Operation of Farm Irrigation Systems, American Society of Agricultural Engineers Monograph 3, M. E. Jensen, ed. ASAE, St. Joseph, MI. pp. 332-341.

Replogle, J.A. and J.L. Merriam. (1980). Scheduling and Management of Irrigation Water Delivery Systems. Proceedings of the American Society of Agricultural Engineers Second National Irrigation Symposium, Oct 20-23. Lincoln, NE.

USDA Natural Resources Conservation Service. (1997). National Engineering Handbook, Part 652, Irrigation Guide.

Zimbelman, D.D., Editor. (1987). Planning, Operation, Rehabilitation and Automation of Irrigation Water Delivery Systems. Proceedings of a Symposium of the Irrigation and Drainage Division, ASCE, July 28-30, Portland, OR. pp. 18-72.

# NOTATION

| | |
|---|---|
| AE | Application efficiency |
| $AE_{lq}$ | Application efficiency, low quarter |
| AR | Advance ratio |
| B/C | Benefit cost ratio |
| CDS | Controlled droplet size |
| CRF | Capital recovery factor |
| CV | Coefficient of variation |
| $D_{avg}$ | Average depth of water accumulated in all elements |
| $d_{lq}$ | Average low quarter depth |
| DU | Distribution uniformity |
| $DU_{low\ half}$ | Distribution uniformity, low half |
| $DU_{lq}$ | Distribution uniformity, low quarter |
| EAE | Equivalent annualized cost factor of escalating energy |
| EAL | Equivalent annual labor |
| ET | Evapotranspiration |
| i | Annual interest rate |
| IE | Irrigation efficiency |
| IS | Irrigation sagacity |
| ITRC | Irrigation Training and Research Center |
| LEPA | Low energy precision application |
| LESA | Low elevation spray application |
| L | Length dimension |
| LQ | lower quarter |
| MAD | Management allowable deficit |
| n | Economic life of a component |
| $PAE_{lq}$ | Potential application efficiency, low quarter |
| SMD | Soil moisture deficit |
| T | Time dimension |
| $T_a$ | Time of application |
| $T_{adv}$ | Time of advance |
| $T_l$ | Time of infiltration at lower end |
| $T_r$ | Time of recession |
| TDH | Total dynamic head |
| UC | Christiansen uniformity coefficient |

# INDEX

Entries are filed word-by-word. Locators in *italics* followed by the notation *eq* indicate equations found in the text, those followed by the notation *fig* indicate figures, those followed by the notation *ph* indicate photos and those followed by the notation *tab* indicate tables.

$AAE_{lq}$ (actual application efficiency) 53
above ground drip irrigation 69–71
abrasion. *See* plugging and abrasion
advance ratio (AR) 29, *29eq*, 30, 39, 40, 42, 43
AE (application efficiency) 2–3, *2eq*
$AE_{lq}$ (application efficiency - low quarter) 4, *107tab*
alfalfa 34, 61, 105
alternate row irrigation 40, *41ph*, 51, 64
application; depth and duration 40, 103; efficiency 34; efficiency, actual ($AAE_{lq}$) 53; efficiency (AE) 2–3, *2eq*; efficiency, low quarter application ($AE_{lq}$) 4, *107tab*; rate 37, 86–89, 94, 95, 98, 99; set time 40; uniformity 85, 88, 98
AR (advance ratio). *See* advance ratio (AR)
asparagus 72
avocados 64

B/C (benefit/cost ratio) 23
basin irrigation 27–32, *28ph*
basin paddy irrigation 36–38, *38ph*
basin systems, design 31–32
bedded irrigation 29–32, *31ph*
beets 29
benefit/cost envelope, economic analysis *8fig*
benefit/cost ratio (B/C) 23
bloom delay 15, 104
border strip irrigation 32–36, *33ph*
buried drip systems 67, *68ph*, 71–73

cablegation 43, *44fig*

California 45, 46
capital costs; annualized 22; construction costs 55, *56–57*; contour ditches (wild flood) irrigation 49; drip/micro irrigation 81–82; equipment, automated 55; return flow systems 55; sprinkler irrigation 110, *110tab*; subirrigation 115, 116; surface irrigation 55–57
capital recovery factor (CRF) *22eq*
cauliflower 72
celery 70
center pivot sprinkler system 95–97, *97ph*
chemigation 61
Christiansen Uniformity Coefficient (UC) 1, 3
citrus 64, 69
climate, irrigation method selection 14
climate modification; bloom delay 15, 104; cooling 69, 101, 104, 114; drip/micro irrigation 64; frost protection 51, 69, 101, 104, 114; humidity 106; irrigation caused 15
coefficient of variation (CV) *3eq*
contaminants, water 13
continuous flood irrigation 36–38, *38ph*
contour ditches irrigation 49
cooling, evaporative 69, 101, 104, 114
corn 29, 73, 91, 105
corrugation irrigation 48
costs; construction 55, *56–57*; design and development 81, 115; drip/micro irrigation *82tab*; energy 58, 76, 82, 104, 110, 116; fertilization 82; initial 76, 81–82, 90, 92, 97, 102; irrigation method selection 23; management 83, 111, 116; micro

systems vs. drip 69; sprinkler systems, solid set 100; subirrigation 116
costs, capital. *See* capital costs
costs, labor. *See* labor costs
costs, operation and maintenance. *See* operation and maintenance costs
CRF (capital recovery factor) *22eq*
crop rotation 71
crops, row 29, 34. *See also* specific crops, e.g. corn
crusting, soil 34, 39
CV (coefficient of variation) *3eq*

de-silting, reservoir 47
debris, floating 13
defoliation 106
dikes 32, 36, 37
distribution uniformity (DU); bedded irrigation 32; border strip irrigation 34; drip/micro irrigation 74, 78; $DU_{low\ half}$ 3; hand move portable/ lateral move portable 91; low quarter ($DU_{lq}$) *3eq*, 53; sprinkler irrigation 107, 108
drainage 38
drip irrigation. *See* drip/micro irrigation; row crow drip irrigation
drip/micro irrigation *25tab*, 61–83, *62fig*, *63fig*
drip tape 70–72
DU (distribution uniformity). *See* distribution uniformity (DU)

EAE (equivalent annualized cost factor of escalating energy) *22eq*, 23
(EAL) equivalent annual labor 23
economic; analysis, irrigation method selection 21–25; efficiency, irrigation development 7–8; returns 23
efficiency, application (AE) 2–3, *2eq*
efficiency, application, low quarter ($AE_{lq}$) 4, *107tab*
efficiency, irrigation methods; cablegation 43; drip/micro irrigation 78; sprinkler irrigation 107–108; subirrigation 114; surface irrigation 52–53; undertree orchard sprinkler system 103
end-tow lateral sprinkler system 91–92, *91ph*
energy availability and reliability, irrigation method selection 15
energy requirements, sprinkler irrigation systems *111tab*
equivalent annual labor (EAL) 23
equivalent annualized cost factor of escalating energy (EAE) *22eq*, 23
erosion control 30
ET (evapotranspiration) 65, 77, 114
evaporation; drip/micro irrigation 75, 78, 79; irrigation scheduling 107; LEPA (Low Energy Precision Application) 98; solid set sprinkler systems 101; sprinkler irrigation 107; subirrigation 114; surface irrigation 52
evapotranspiration (ET) 65, 77, 114

failure, catastrophic 67
fertigation 64, 72, *75ph*
fertilization 74, 82
fill and drain irrigation 38–39
filtration. *See also* plugging and abrasion; above ground drip 70; center pivot sprinkler system 97; drip/micro irrigation 61, 77; media in row crop drip *73ph*; micro systems vs. drip 69; permanent drip 72; row crop drip (subsurface) *73ph*; sprinkler irrigation 104, 106
flat-planted irrigation 29
flow rates 48, *64tab*, 70
frost protection 51, 69, 101, 104, 114
furrow irrigation 39–48

grading. *See also* leveling; border strip irrigation 35; continuous flood (basin paddy) irrigation 37; drip/micro irrigation 74; furrow irrigation 47–48; laser controlled 31, 37, 47; requirements 48, 104;

sprinkler irrigation 105; surface irrigation 54

hand move portable sprinkler system 89–91, *90ph*
hay 34, 76, 91, 105

IE (irrigation efficiency) 1–2, *1eq*
infiltration 30–31, 41
intake rate 105; furrow irrigation 41, 47; low pressure spray nozzles 88; sprinkler irrigation 105; surface irrigation 50, 53; surge flow irrigation 45
irrigation efficiency (IE) 1–2, *1eq*
irrigation methods; above ground drip 69–71; alternate row 40, *41ph*, 51, 64; basin 27–32, *28ph*; basin paddy 36–38; bedded 29–32, *31ph*; border strip 32–36, *33ph*; continuous flood 36–38, *38ph*; contour ditches 49; corrugation 48; drip/micro 61–83; fill and drain 38–39; flat-planted 29; furrow 39–48; moving water 27; orchard/vineyard drip 65–67, *66ph*; orchard/vineyard micro irrigation 68–70; ponded water 27–32, *28ph*, 38–39; row crop drip *63ph*, 69–73, *73ph*; sprinkler 85–112; sub-irrigation 113–117; surface 27–60, *28tab*; surge flow 44–46, *45fig*, *46ph*; trickle. *See* drip/micro irrigation; wild flood 49
irrigation sagacity (IS) *2eq*
irrigation systems; design 86; developing countries 10, *11tab*; labor requirements. *See* labor requirements; performance 1, 4, 82, 83; selection guide *16–19*
Irrigation Training and Research Center (ITRC) 4
IS (irrigation sagacity) *2eq*
ITRC (Irrigation Training and Research Center) 4

labor costs; drip/micro irrigation 82–83; furrow irrigation 42–43; hand move portable/lateral move portable sprinkler system 90; sprinkler irrigation 111; sub-irrigation 116; surface irrigation 58–59
labor requirements; basin irrigation 30; border strip irrigation 34; cablegation 44; center pivot sprinkler system 97; contour ditches (wild flood) irrigation 49; corrugation 48; drip/micro irrigation *25tab*, 79–80; furrow irrigation 48; solid set systems 100, 102; sprinkler irrigation *25tab*, 104, 109, *109tab*; subirrigation 115; surface irrigation 54, *58tab*; wheel move systems 94
lateral move portable sprinkler system 89–91
lateral move sprinkler system 98–99
leaching 13, 39, 51, 76. *See also* salinity; water quality
legal issues 10
LEPA (low energy precision application) 98–99, 107
LESA (low elevation spray application) 98–99, 107
lettuce 61, 71, 72
leveling 31, 37. *See also* grading
linear move sprinkler system 98–99, *98ph*, *99ph*, *100ph*
low elevation spray application (LESA) 98–99, 107
low energy precision application (LEPA) 98–99, 107
low pressure spray nozzles 87–88

MAD (management allowed deficit) 4, 34, 35, 40, 42
maintenance costs. *See* operation and maintenance costs
management allowed deficit (MAD) 4, 34, 35, 40, 42

management skills 12, 53, 72, 79, 115
manganese 78
melons 29, 39, 77
microsystems 62ph, 68–70, 69ph, 73–75. *See also* drip/micro irrigation
mold 71
moving water irrigation 27

nematode control 113
nozzles 87–88
nutrient uptake 74

obstructions, physical 14, 97
onions 71, 99, 105
operation and maintenance costs; above ground drip 70; annualized 2; center pivot systems 97; drip/micro irrigation 80, 81, 83; sprinkler irrigation 111, *112tab*; subirrigation 116; surface irrigation 55t, 59, *60tab*; system components *24tab*
orchard; crops 29, 39, 64, *68ph*, 81, 97; overtree sprinklers 103–104; undertree sprinklers 88
orchard/vineyard; drip irrigation 65–67, *66ph*; micro irrigation 68–70
over-irrigation 74

$PAE_{lq}$ (potential application efficiency, low quarter) 3–4, *3eq*, 52
pasture (crop) 34, 76
peas 70
peppers 61, 71, 76
perforated pipe 88–89
performance, irrigation methods 1, 4, 82, 83
physical conditions, irrigation method selection 12–19, *15tab*
pipe, perforated 88–89
plugging and abrasion 13, 66, 69, 78. *See also* filtration
political issues, irrigation method selection 10

ponded water irrigation 38–39
ponding, surface 34, 105
portable sprinkler systems. *See* sprinkler systems, portable
potatoes 29, 70, 99
potential application efficiency, low quarter ($PAE_{lq}$) 3–4, *3eq*, 52
potential application efficiency, surface irrigation *52fig*

rainfall storage 51
reservoirs 34, 47, 77
return flow systems; capital costs 55; furrow irrigation 42; tailwater 13, 27, 39, 43, 45, 55; underground distribution *42fig*
rice 36
risk assessment 26
root intrusion 70, 72
rot, trunk and root 76
rotating head; impact sprinklers 87, 89; sprinklers 85–87, *96ph*
rotator sprinklers 88
row crop drip irrigation; above ground 69–71; subsurface *63ph*, 71–73, *73ph*
runoff 43, 74, 79, 107

salinity. *See also* water quality; drip/micro irrigation 74, 77–78; salt accumulation 14; sprinkler irrigation 105; subirrigation 114; surface irrigation 51; undertree orchard sprinkler system 103
scheduling, irrigation 53–54, 78–79, 104, 108
seepage 114
side move lateral sprinkler system 93–94, *94ph*
side roll lateral sprinkler system 92
side roll lateral/wheel line sprinkler system 92–93, *93ph*
site conditions, irrigation method selection 9–21
SMD (soil-moisture deficit) 4, 34, 53, 54, 108

soil; bearing strength 97; clay 77; coarse grained 113; coral sand 77; cracking 36; drip/micro irrigation 77; infiltration, surface irrigation 27; intake rate 41, 53, 54, 95; intake rate, continuous flood irrigation 36; intake rate, spray nozzles 88; intake rate, surface irrigation 47–48, 50; moisture-deficit (SMD). *See* SMD (soil moisture-deficit); sandy 45; shallow 35; sticky 91
solid set drip/micro systems 61, 64
solid set sprinkler systems 100–102, *101ph*
sprinkler systems; advantages and disadvantages 104–105; center pivot 95–97, *97ph*; end-tow lateral 91–92, *91ph*; energy requirements *111tab*; irrigation 85–112; labor requirements *25tab*; lateral move *98ph*, 99, *99ph*, *100ph*; linear move *98ph*, 99, *99ph*, *100ph*; overtree orchard 103–104; portable 89–91; side move lateral 93–94, *94ph*; side roll lateral/wheel line 92–93, *93ph*; solid set 100–102, *101ph*; traveling gun and rotating boom 94–95; undertree orchard 88, 102–103, *102ph*
strawberries 71, 72
stream size 32, 34, 40, 44
subirrigation 113–117
sugar cane 71, 72
surface irrigation 27–60, *28tab*
surge flow irrigation 44–46, *45fig*, *46ph*

tailwater return flow systems 13, 27, 39, 43, 45, 55
Texas 45
tile water 13
tomatoes 72, 76, 105
topography; border strip irrigation 35; continuous flood irrigation 37; drip/micro irrigation 74, 77; irregular 4, 48, 50, 88, 106;
irrigation method selection 14; portable sprinkler systems 91; sprinkler irrigation 105–106; subirrigation 114; surface irrigation 50
training, labor 4, 48, 54, 80, 109
traveling gun and rotating boom sprinkler system 94–95, *96ph*
trickle irrigation. *See* drip/micro irrigation

UC (Christiansen Uniformity Coefficient) 1, 3
under irrigation 3, 79, 80
undertree sprinklers 88, 102–103, *102ph*
uniformity coefficient (UC) 1, 3
uniformity, distribution (DU). *See* distribution uniformity (DU)

vineyards 29, 34, 39, 65, 81, 97. *See also* orchard/vineyard

water; application rate 37, 86–89, 94, 95, 98, 99; cost savings 79; delivery 5, 36, 47, 55, 76; distribution 44; gated pipe application 42; quality 13–14, 61, 77, 106; reuse waters 13, 27, 39, 43, 45, 55
water supply; basin irrigation 30; constant rate 106; drip/micro irrigation 77; fixed 42; flexible 77; irrigation method selection 13; rotation schedule 43; sprinkler irrigation 106; subirrigation 114; surface irrigation 51; variable stream 39
water table 14, 113
wheat 72
wild flood irrigation 49
wind; drift 85; drip/micro irrigation 78; low pressure spray nozzles 88; side roll/wheel line sprinkler system 93; sprinkler irrigation 107; subirrigation 114; surface irrigation 52

yield 43, 80–81, 82